Series in Mathematical Analysis and Applications
Edited by Ravi P. Agarwal and Donal O'Regan

VOLUME 10

TOPOLOGICAL DEGREE THEORY AND APPLICATIONS

SERIES IN MATHEMATICAL ANALYSIS AND APPLICATIONS

Series in Mathematical Analysis and Applications (SIMAA) is edited by Ravi P. Agarwal, Florida Institute of Technology, USA and Donal O'Regan, National University of Ireland, Galway, Ireland.

The series is aimed at reporting on new developments in mathematical analysis and applications of a high standard and or current interest. Each volume in the series is devoted to a topic in analysis that has been applied, or is potentially applicable, to the solutions of scientific, engineering and social problems.

Volume 1
Method of Variation of Parameters for Dynamic Systems
V. Lakshmikantham and S.G. Deo

Volume 2
Integral and Integrodifferential Equations: Theory, Methods and Applications
Edited by Ravi P. Agarwal and Donal O'Regan

Volume 3
Theorems of Leray-Schauder Type and Applications
Donal O'Regan and Radu Precup

Volume 4
Set Valued Mappings with Applications in Nonlinear Analysis
Edited by Ravi P. Agarwal and Donal O'Regan

Volume 5
Oscillation Theory for Second Order Dynamic Equations
Ravi P. Agarwal, Said R. Grace, and Donal O'Regan

Volume 6
Theory of Fuzzy Differential Equations and Inclusions
V. Lakshmikantham and Ram N. Mohapatra

Volume 7
Monotone Flows and Rapid Convergence for Nonlinear Partial Differential Equations
V. Lakshmikantham, S. Koksal, and Raymond Bonnett

Volume 8
Nonsmooth Critical Point Theory and Nonlinear Boundary Value Problems
Leszek Gasiński and Nikolaos S. Papageorgiou

Volume 9
Nonlinear Analysis
Leszek Gasiński and Nikolaos S. Papageorgiou

Volume 10
Topological Degree Theory and Applications
Donal O'Regan, Yeol Je Cho, and Yu-Qing Chen

Series in Mathematical Analysis and Applications

Edited by Ravi P. Agarwal and Donal O'Regan

VOLUME 10

TOPOLOGICAL DEGREE THEORY AND APPLICATIONS

Donal O'Regan

Yeol Je Cho

Yu-Qing Chen

Chapman & Hall/CRC

Taylor & Francis Group

Boca Raton London New York

Published in 2006 by
Chapman & Hall/CRC
Taylor & Francis Group
6000 Broken Sound Parkway NW, Suite 300
Boca Raton, FL 33487-2742

Printed in the United States of America on acid-free paper
10 9 8 7 6 5 4 3 2 1

International Standard Book Number-10: 1-58488-648-X (Hardcover)
International Standard Book Number-13: 978-1-58488-648-8 (Hardcover)

Library of Congress Cataloging-in-Publication Data

Catalog record is available from the Library of Congress

informa
Taylor & Francis Group
is the Academic Division of Informa plc.

Visit the Taylor & Francis Web site at
http://www.taylorandfrancis.com

and the CRC Press Web site at
http://www.crcpress.com

Contents

Chapter 1

BROUWER DEGREE THEORY

Let R be the real numbers, $R^n = \{x = (x_1, x_2, \cdots, x_n) : x_i \in R \text{ for } i = 1, 2, \cdots, n\}$ with $|x| = (\sum_{i=1}^{n} x_i^2)^{\frac{1}{2}}$ and let $\Omega \subset R^n$, and let $f : \Omega \to R^n$ be a continuous function. A basic mathematical problem is: Does $f(x) = 0$ have a solution in Ω? It is also of interest to know how many solutions are distributed in Ω. In this chapter, we will present a number, the topological degree of f with respect to Ω and 0, which is very useful in answering these questions. To motivate the process, let us first recall the winding number of plane curves, a basic topic in an elementary course in complex analysis. Let C be the set of complex numbers, $\Gamma \subset C$ an oriented closed C^1 curve and $a \in C \setminus \Gamma$. Then the integer

$$w(\Gamma, a) = \frac{1}{2\pi i} \int_{\Gamma} \frac{1}{z - a} dz \tag{1}$$

is called the winding number of Γ with respect to $a \in C \setminus \Gamma$. Now, let $G \subset C$ be a simply connected region and $f : G \to C$ be analytic and $\Gamma \subset G$ a closed C^1 curve such that $f(z) \neq 0$ on Γ. Then we have

$$w(f(\Gamma), 0) = \frac{1}{2\pi i} \int_{f(\Gamma)} \frac{1}{z} dz = \frac{1}{2\pi i} \int_{\Gamma} \frac{f'(z)}{f(z)} dz = \sum_i w(\Gamma, z_i)\alpha_i, \tag{2}$$

where z_i are the zeros of f in the region enclosed by Γ and α_i are the corresponding multiplicites. If we assume in addition that Γ has positive orientation and no intersection points, then we know from Jordan's Theorem, which will be proved later in this chapter, that $w(\Gamma, z_i) = 1$ for all z_i. Thus (2) becomes

$$w(f(\Gamma), 0) = \sum_i \alpha_i. \tag{3}$$

So we may say that f has at least $|w(f(\Gamma), 0)|$ zeros in G. The winding number is a very old concept which goes back to Cauchy and Gauss. Kronecker, Hadamard, Poincare, and others extended formula (1). In 1912, Brouwer [32] introduced the so-called Brouwer degree in R^n (see Browder [35], Sieberg [277] for historical developments). In this chapter, we introduce the Brouwer degree theory and its generalization to functions in VMO. This chapter is organized as follows:

In Section 1.1 we introduce the notion of a critical point for a differentiable function f. We then prove Sard's Lemma, which states that the set of critical points of a C^1 function is "small". Our final result in this section shows how a continuous function can be approximated by a C^∞ function.

In Section 1.2 we begin by defining the degree of a C^1 function using the Jacobian. Also we present an integral representation which we use to define the degree of a continuous function. Also in this section we present some properties of our degree (see theorems 1.2.6, 1.2.12, and 1.2.13) and some useful consequences. For example, we prove Brouwer's and Borsuk's fixed point theorem, Jordan's separation theorem and an open mapping theorem. In addition we discuss the relation between the winding number and the degree.

In Section 1.3 we discuss some properties of the average value function and then we introduce the degree for functions in VMO.

In Section 1.4 we use the degree theory in Section 1.2 to present some existence results for the periodic and anti-periodic first order ordinary differential equations.

1.1 Continuous and Differentiable Functions

We begin with the following Bolzano's intermediate value theorem:

Theorem 1.1.1. Let $f : [a, b] \to R$ be a continuous function, then, for m between $f(a)$ and $f(b)$, there exists $x_0 \in [a, b]$ such that $f(x_0) = m$.

Corollary 1.1.2. Let $f : [a, b] \to R$ be a continuous function such that $f(a)f(b) < 0$. Then there exists $x_0 \in (a, b)$ such that $f(x_0) = 0$.

Corollary 1.1.3. Let $f : [a, b] \to [a, b]$ be a continuous function. Then there exists $x_0 \in [a, b]$ such that $f(x_0) = x_0$.

Let $\Omega \subset R^n$ be an open subset. We recall that a function $f : \Omega \to R^n$ is *differentiable* at $x_0 \in \Omega$ if there is a matrix $f'(x_0)$ such that $f(x_0 + h) = f(x_0) + f'(x_0)h + o(h)$, where $x_0 + h \in \Omega$ and $\frac{|o(h)|}{|h|}$ tends to zero as $|h| \to 0$.

We use $C^k(\Omega)$ to denote the space of k-times continuously differentiable functions. If f is differentiable at x_0, we call $J_f(x_0) = det f'(x_0)$ the Jacobian of f at x_0. If $J_f(x_0) = 0$, then x_0 is said to be a critical point of f and we use $S_f(\Omega) = \{x \in \Omega : J_f(x) = 0\}$ to denote the set of critical points of f, in Ω. If $f^{-1}(y) \cap S_f(\Omega) = \emptyset$, then y is said to be a regular value of f. Otherwise, y is said to be a singular value of f.

Lemma 1.1.4. (Sard's Lemma) Let $\Omega \subset R^n$ be open and $f \in C^1(\Omega)$. Then $\mu_n(f(S_f(\Omega))) = 0$, where μ_n is the n-dimensional Lebesgue measure.

Proof. Since Ω is open, $\Omega = \cup_{i=1}^{\infty} Q_i$, where Q_i is a cube for $i = 1, 2, \cdots$. We only need to show that $\mu_n(f(S_f(Q))) = 0$ for a cube $Q \subset \Omega$. In fact, let

l be the lateral length of Q. By the uniform continuity of f' on Q, for any given $\epsilon > 0$, there exists an integer $m > 0$ such that

$$|f'(x) - f'(y)| \leq \epsilon$$

for all $x, y \in Q$ with $|x - y| \leq \frac{\sqrt{n}l}{m}$. Therefore, we have

$$\begin{aligned}|f(x) - f(y) - f'(y)(x - y)| &\leq \int_0^1 |f'(y + t(x - y)) - f'(y)||x - y|dt \\ &\leq \epsilon|x - y|\end{aligned}$$

for all $x, y \in Q$ with $|x - y| \leq \frac{\sqrt{n}l}{m}$. We decompose Q into r cubes, Q^i, of diameter $\frac{\sqrt{n}l}{m}$, $i = 1, 2, \cdots, r$. Since $\frac{l}{m}$ is the lateral length of Q^i, we have $r = m^n$. Now, suppose that $Q^i \cap S_f(\Omega) \neq \emptyset$. Choosing $y \in Q^i \cap S_f(\Omega)$, we have $f(y+x) - f(y) = f'(y)x + R(y, x)$ for all $x \in Q^i - y$, where $|R(y, x+y)| \leq \epsilon \frac{\sqrt{n}l}{m}$. Therefore, we have

$$f(Q^i) = f(y) + f'(y)(Q^i - y) + R(y, Q^i).$$

But $f'(y) = 0$, so $f'(y)(Q^i - y)$ is contained in an $(n-1)$- dimensional subspace of R^n. Thus, $\mu_n(f'(y)(Q^i - y)) = 0$, so we have

$$\mu_n(f(Q^i)) \leq 2^n \epsilon^n (\frac{\sqrt{n}l}{m})^n.$$

Obviously, $f(S_f(Q)) \subset \cup_{i=1}^r f(Q^i)$, so we have

$$\mu_n(f(S_f(Q)) \leq r 2^n \epsilon^n (\frac{\sqrt{n}l}{m})^n = 2^n \epsilon^n (\sqrt{n}l)^n.$$

By letting $\epsilon \to 0^+$, we obtain $\mu_n(f(S_f(Q))) = 0$. Therefore, $\mu_n(f(S_f(\Omega))) = 0$. This completes the proof.

Proposition 1.1.5. Let $K \subset R^n$ be a bounded closed subset, and $f : K \to R^n$ continuous. Then there exists a continuous function $\tilde{f} : R^n \to \overline{conv f(K)}$ such that $\tilde{f}(x) = f(x)$ for all $x \in K$, where $conv f(K)$ is the convex hull of $f(K)$.

Proof. Since K is bounded closed subset, there exists at most countable $\{k_i : i = 1, 2, \cdots\} \subset K$ such that $\overline{\{k_i : i = 1, 2, \cdots\}} = K$. Put

$$d(x, K) = \inf_{y \in K} |x - y|, \quad \alpha_i(x) = \max\{2 - \frac{|x - k_i|}{d(x, A)}, 0\}$$

for any $x \notin K$ and

$$\tilde{f}(x) = \begin{cases} f(x), & x \in K, \\ \frac{\sum_{i \geq 1} 2^{-i} \alpha_i(x) f(k_i)}{\sum_{i \geq 1} 2^{-i} \alpha_i(x)}, & x \notin K. \end{cases}$$

Then $\tilde{f}$ is the desired function.

Proposition 1.1.6. Let $K \subset R^n$ be a bounded closed subset and $f : K \to R^n$ continuous. Then there exists a function $g \in C^\infty(R^n)$ such that $|f(x) - g(x)| < \epsilon$.

Proof. By Proposition 1.1.5, there exists a continuous extension $\tilde{f}$ of f to R^n. Define the following function

$$\phi(x) = \begin{cases} ce^{-\frac{1}{1-|x|}}, & |x| < 1, \\ 0, & |x| \geq 1, \end{cases} \tag{1.1}$$

where c satisfies $\int_{R^n} \phi(x)dx = 1$. Set $\phi_\lambda(x) = \lambda^{-n}\phi(\frac{x}{\lambda})$ for all $x \in R^n$ and

$$f_\lambda(x) = \int_{R^n} \tilde{f}(y)\phi_\lambda(y-x)dx \text{ for all } x \in R^n,\ \lambda > 0.$$

It is obvious that $suppf_\lambda = \overline{\{x \in R^n : f_\lambda(x) \neq 0\}} = \{x : |x| \leq \lambda\}$ for all $\lambda > 0$. Consequently, we have $f_\lambda \in C^\infty$ and $f_\lambda(x) \to f(x)$ uniformly on K as $\lambda \to 0^+$. Taking g as f_λ for sufficiently small λ, g is the desired function. This completes the proof.

1.2 Construction of Brouwer Degree

Now, we give the construction of Brouwer degree in this section as follows:

Definition 1.2.1. Let $\Omega \subset R^N$ be open and bounded and $f \in C^1(\overline{\Omega})$. If $p \notin f(\partial\Omega)$ and $J_f(p) \neq 0$, then we define

$$deg(f, \Omega, p) = \sum_{x \in f^{-1}(p)} sgnJ_f(x),$$

where $deg(f, \Omega, p) = 0$ if $f^{-1}(p) = \emptyset$.

The next result gives another equivalent form of Definition 1.2.1.

Proposition 1.2.2. Let Ω, f and p be as in Definition 1.2.1 and let

$$\phi_\epsilon(x) = \begin{cases} c\epsilon^{-n}e^{-\frac{1}{1-|\epsilon^{-1}x|^2}}, & |x| < 1, \\ 0, & \text{otherwise}, \end{cases} \tag{1.2}$$

where c is a constant such that $\int_{R^n} \phi(x) = 1$. Then there exists $\epsilon_0 = \epsilon_0(p, f)$ such that

$$deg(f, \Omega, p) = \int_\Omega \phi_\epsilon(f(x) - p)J_f(x)dx \quad \text{for all } \epsilon \in (0, \epsilon_0).$$

Proof. The case $f^{-1}(p) = \emptyset$ is obvious. Assume that

$$f^{-1}(p) = \{x_1, x_2, \cdots, x_n\}.$$

We can find disjoint balls $B_r(x_i)$ and a neighborhood V_i of p such that $f : B_r(x_i) \to V_i$ is a homeomorphism and $sgnJ_f(x) = sgnJ_f(x_i)$ in $B_i(x_i)$. We may take $r_0 > 0$ such that $B_{r_0}(p) \subset \cap_{i=1}^n V_i$ and set $U_i = B_r(x_i) \cap f^{-1}(B_{r_0}(p))$. Then $|f(x) - p| \geq \delta$ on $\overline{\Omega} \setminus \cup_{i=1}^n U_i$ for some $\delta > 0$ and so, for any $\epsilon < \delta$, we have

$$\int_\Omega \phi_\epsilon(f(x) - p) J_f(x) dx = \sum_{i=1}^n sgnJ_f(x_i) \int_{U_i} \phi_\epsilon(f(x) - p)|J_f(x)| dx.$$

But we have

$$\begin{aligned} & J_f(x) = J_{f-p}(x), \\ & \int_{U_i} \phi_\epsilon(f(x) - p)|J_f(x)| dx = \int_{B_{r_0}} \phi_\epsilon(x) dx = 1, \\ & \epsilon < \min\{r_0, \delta\}. \end{aligned}$$

This completes the proof.

Definition 1.2.3. Let $\Omega \subset R^N$ be open and bounded and $f \in C^2(\overline{\Omega})$. If $p \notin f(\partial\Omega)$. Then we define

$$deg(f, \Omega, p) = deg(f, \Omega, p'),$$

where p' is any regular value of f that $|p' - p| < d(p, f(\partial\Omega))$.

We need to check that, for any two regular values p_1 and p_2 of f,

$$deg(f, \Omega, p_1) = deg(f, \Omega, p_2).$$

For any $\epsilon < d(p, f(\partial\Omega)) - \max\{|p - p_i| : i = 1, 2\}$, we have

$$deg(f, \Omega, p_i) = \int_\Omega \phi_\epsilon(f(x) - p_i) J_f(x) dx \quad \text{for } i = 1, 2.$$

Notice that

$$\phi_\epsilon(x - p_2) - \phi_\epsilon(x - p_1) = divw(x),$$

where

$$w(x) = (p_1 - p_2) \int_0^1 \phi_\epsilon(x - p_1 + t(p_1 - p_2)) dt.$$

We show that there exists a function $v \in C^1(R^N)$ such that $supp(v) \subset \Omega$ and

$$[\phi_\epsilon(f(x) - p_2) - \phi_\epsilon(f(x) - p_1)] J_f(x) = divv(x) \text{ for all } x \in \Omega.$$

Lemma 1.2.4. Let $\Omega \subset R^N$ be open, $f \in C^2(\overline{\Omega})$ and let d_{ij} be the cofactor of $\frac{\partial f_j}{\partial x_i}$ in $J_f(x)$ and

$$v_i(x) = \begin{cases} \sum_{j=1}^N w_j(f(x))d_{ij}(x) & x \in \overline{\Omega}, \\ 0, & \text{otherwise}. \end{cases}$$

Then $(v_1(x), v_1(x), \cdots, v_N(x))$ satisfies $divv(x) = divw(f(x))J_f(x)$.

Proof. Since $supp(w) \subset \overline{B(p,r)}$ for $r \leq max\{|p - p_i| : i = 1, 2\} + \epsilon < d(p, \partial\Omega)$, we have

$$supp(v) \subset \Omega,$$

$$\partial_i v_i(x) = \sum_{j,k=1}^N d_{jk}\partial_k W_j(f(x))\partial_i f_k(x) + \sum_{j=1}^N W_j(f(x))\partial_i d_{ij}(x),$$

where $\partial_k = \frac{\partial}{\partial x_k}$. Now, we claim that

$$\sum_{i=1}^{\ \ N} \partial_i d_{ij}(x) = 0 \quad \text{for } j = 1, 2, \cdots, N.$$

For any given j, let f_{x_k} denote the column

$$(\partial_k f_1, \cdots, \partial_k f_{j-1}, \partial_k f_{j+1}, \cdots, \partial_k f_n).$$

Then we have

$$d_{ij}(x) = (-1)^{i+j} det(f_{x_1}, \cdots, f_{i-1}, f_{i+1}, \cdots, f_N).$$

Therefore, it follows that

$$\partial_i d_{ij}(x) = (-1)^{i+j} \sum_{k=1}^N det(f_{x_1}, \cdots, f_{x_{i-1}}, f_{x_{i+1}}, \cdots, \partial_i f_{x_k}, \cdots, f_{x_N}).$$

Set

$$a_{ki} = det(\partial_i f_{x_k}, f_{x_1}, \cdots, f_{x_{i-1}}, f_{x_{i+1}}, \cdots, f_{x_{k-1}}, f_{x_{k+1}}, \cdots, f_{x_N}),$$

then we have $a_{ki} = a_{ik}$ and

$$\begin{aligned}(-1)^{i+j}\partial_i d_{ij}(x) &= \sum_{i,k=1}^N (-1)^{k-1} a_{ki} + \sum_{k>i} (-1)^{k-2} a_{ki} \\ &= \sum_{k=1}^N (-1)^{k-1} \delta_{ki} a_{ki},\end{aligned}$$

where $\delta_{ki} = 1$ for $k < i$, $\delta_{ii} = 0$ and $\delta_{ki} = -\delta_{ik}$ for $i, k = 1, 2, \cdots, N$. Hence we have

$$(-1)^j \sum_{i=1}^{N} \partial_i d_{ij}(x) = \sum_{i,k=1}^{N} (-1)^{k-1+i} \gamma_{ki} a_{ki} = \sum_{k,i=1}^{N} (-1)^{i-1+k} \gamma_{ik} a_{ik}$$
$$= -\sum_{i,k=1}^{N} (-1)^{k-1+i} \gamma_{ki} a_{ki} = 0.$$

Now, we have

$$\partial_i v_i(x) = \sum_{j,k=1}^{N} d_{i,j} \partial_k w_j(f(x)) \partial_i f_k(x) + \sum_{j=1}^{N} w_j(f(x)) \partial_i d_{ij}(x).$$

On the other hand, $\sum_{i=1}^N d_{ij} \partial_i f_k(x) = \delta_{jk} J_f(x)$ with Kronecker's δ_{jk}. Therefore, it follows that

$$divv(x) = \sum_{k,j=1}^{N} \partial_k w_j(f(x)) \delta_{jk} J_f(x) = divw(f(x)) J_f(x).$$

This completes the proof.

Finally, we are ready to introduce the following definition:

Definition 1.2.5. Let $\Omega \subset R^N$ be open and bounded, $f \in C(\overline{\Omega})$ and $p \notin f(\partial\Omega)$. Then we define

$$deg(f, \Omega, p) = deg(g, \Omega, p),$$

where $g \in C^2(\overline{\Omega})$ and $|g - f| < d(p, f(\partial\Omega))$.

Now, one may check the following properties by a reduction to the regular case.

Theorem 1.2.6. Let $\Omega \subset R^N$ be an open bounded subset and $f : \overline{\Omega} \to R^N$ be a continuous mapping. If $p \notin f(\partial\Omega)$, then there exists an integer $deg(f, \Omega, p)$ satisfying the following properties:

(1) (*Normality*) $deg(I, \Omega, p) = 1$ if and only if $p \in \Omega$, where I denotes the identity mapping;

(2) (*Solvability*) If $deg(f, \Omega, p) \neq 0$, then $f(x) = p$ has a solution in Ω;

(3) (*Homotopy*) If $f_t(x) : [0,1] \times \overline{\Omega} \to R^N$ is continuous and $p \notin \cup_{t \in [0,1]} f_t(\partial\Omega)$, then $deg(f_t, \Omega, p)$ does not depend on $t \in [0,1]$;

(4) (*Additivity*) Suppose that Ω_1, Ω_2 are two disjoint open subsets of Ω and $p \notin f(\overline{\Omega} - \Omega_1 \cup \Omega_2)$. Then $deg(f, \Omega, p) = deg(f, \Omega_1, p) + deg(f, \Omega_2, p)$;

(5) $deg(f, \Omega, p)$ is a constant on any connected component of $R^n \setminus f(\partial\Omega)$.

As consequences of Theorem 1.2.6, we have the following results:

Theorem 1.2.7. Let $f : \overline{B(0,R)} \subset R^n \to \overline{B(0,R)}$ be a continuous mapping. If $|f(x)| \leq R$ for all $x \in \partial B(0,R)$, then f has a fixed point in $\overline{B(0,R)}$.

Proof. We may assume that $x \neq f(x)$ for all $x \in \partial B(0,R)$. Put $H(t,x) = x - tf(x)$ for all $(t,x) \in [0,1] \times \overline{B(0,R)}$. Then $0 \neq H(t,x)$ for all $[0,1] \times \partial B(0,R)$. Therefore, we have

$$deg(I - f, B(0,R), 0) = deg(I, B(0,R), 0) = 1.$$

Hence f has a fixed point in $\overline{B(0,R)}$. This completes the proof.

From Theorem 1.2.7, we have the well-known Brouwer fixed point theorem:

Theorem 1.2.8. Let $C \subset R^n$ be a nonempty bounded closed convex subset and $f : C \to C$ be a continuous mapping. Then f has a fixed point in C.

Proof. Take $B(0,R)$ such that $C \subset B(0,R)$ and let $r : \overline{B(0,R)} \to C$ be a retraction. By Theorem 1.2.7, there exists $x_0 \in \overline{B(0,R)}$ such that $frx_0 = x_0$. Therefore, $x_0 \in C$, and so we have $rx_0 = x_0$. This completes the proof.

Theorem 1.2.9. Let $f : R^n \to R^n$ be a continuous mapping and $0 \in \Omega \subset R^n$ with Ω an open bounded subset. If $(f(x), x) > 0$ for all $x \in \partial\Omega$, then $deg(f, \Omega, 0) = 1$.

Proof. Put $H(t,x) = tx + (1-t)f(x)$ for all $(t,x) \in [0,1] \times \overline{\Omega}$. Then $0 \notin H([0,1] \times \partial\Omega)$, and so we have

$$deg(f, \Omega, 0) = deg(I, \Omega, 0) = 1.$$

This completes the proof.

Corollary 1.2.10. Let $f : R^n \to R^n$ be a continuous mapping. If

$$\lim_{|x| \to \infty} \frac{(f(x), x)}{|x|} = +\infty,$$

then $f(R^n) = R^n$.

Proof. For any $p \in R^n$, it is easy to see that there exists $R > 0$ such that $(f(x) - p, x) > 0$ for all $x \in \partial B(0,R)$, where $B(0,R)$ is the open ball centered at zero with radius R. By Theorem 1.2.9, we have

$$deg(f - p, B(0,R), 0) = 1$$

and so $f(x) - p = 0$ has a solution in $B(0,R)$. This completes the proof.

Theorem 1.2.11. (**Borsuk's Theorem**) Let $\Omega \subset R^n$ be open bounded and symmetric with $0 \in \Omega$. If $f \in C(\overline{\Omega})$ is odd and $0 \notin f(\partial\Omega)$, then $d(f, \Omega, 0)$ is odd.

Proof. Without loss of generality, we may assume that $f \in C^1(\overline{\Omega})$ with $J_f(0) \neq 0$. Next, we define a mapping $g \in C^1(\overline{\Omega})$ sufficiently close to f by induction as follows:

Let $\phi \in C^1(R)$ be an odd mapping with $\phi'(0) = 0$ and $\phi(t) = 0$ if and only if $t = 0$. Put $\Omega_k = \{x \in \Omega : x_k \neq 0\}$ and $h(x) = \frac{f(x)}{\phi(x_1)}$ for all $x \in \Omega_1$. Choose $|y_1|$ sufficiently small such that y_1 is a regular value for h on Ω_1. Put $g_1(x) = f(x) - \phi(x_1)y_1$, then 0 is a regular value for g_1 on Ω_1.

Suppose that we have already an odd $g_k \in C^1(\overline{\Omega})$ close to f such that 0 is a regular value for g_k on Ω_k. Then we define $g_{k+1}(x) = g_k(x) - \phi(x_{k+1})y_{k+1}$ with $|y_{k+1}|$ small enough such that 0 is a regular value for g_{k+1} on Ω_{k+1}.

If $x \in \Omega_{k+1}$ and $x_{k+1} = 0$, then

$$x \in \Omega_k, \quad g_{k+1}(x) = g_k(x), \quad g'_{k+1}(x) = g'_k(x)$$

and hence $J_{g_{k+1}}(x) \neq 0$. By induction, we also have $g'_n(0) = g'_1(0) = f'(0)$ and so 0 is a regular value for g_n. By Definition 1.2.5 and Definition 1.2.1, we know that

$$deg(f, \Omega, 0) = deg(g_n, \Omega, 0) = sgnJ_{g_n}(0) + \sum_{x \in g^{-1}(0), x \neq 0} sgnJ_{g_n}(x)$$

and thus $deg(f, \Omega, 0)$ is odd. This completes the proof.

The following theorem shows the relationship between Brouwer degrees in different dimensional spaces:

Theorem 1.2.12. Let $\Omega \subset R^n$ be an open bounded subset, $1 \leq m < n$, let $f : \overline{\Omega} \to R^m$ be a continuous function and let $g = I - f$. If $y \notin (I - f)(\partial\Omega)$, then

$$deg(g, \Omega, y) = deg(g_m, \Omega \cap R^m, y),$$

where g_m is the restriction of g on $\overline{\Omega} \cap R^m$.

Proof. We may assume that $f \in C^2(\overline{\Omega})$ and y is a regular value for g on $\overline{\Omega}$ A direct computation yields that $J_g(x) = J_{g_m}(x)$ and so the conclusion follows from Definition 1.2.1. This completes the proof.

Let $\Omega \subset R^n$ be open and bounded and let $f \in C(\overline{\Omega})$. By the homotopy invariance of $deg(f, \Omega, y)$, we know that $deg(f, \Omega, y)$ is the same integer as y ranges through the same connected component U of $R^n \setminus f(\partial\Omega)$. Therefore, it is reasonable to denote this integer by $deg(f, \Omega, U)$. The unbounded connected component is denoted by U_∞. Now, we have the product formula:

Theorem 1.2.13. Let $\Omega \subset R^n$ be an open bounded subset, $f \in C(\overline{\Omega})$, $g \in C(R^n)$ and let U_i be the bounded connected components of $R^n \setminus f(\partial\Omega)$.

If $p \notin (gf)(\partial\Omega)$, then

$$deg(gf, \Omega, p) = \sum_i deg(f, \Omega, U_i) deg(g, U_i, p), \qquad (1.2.1)$$

where only finitely many terms are not zero.

Proof. We first prove (1.2.1) only has finitely many non-zero terms. Take $r > 0$ such that $f(\overline{\Omega}) \subset B_r(0)$. Then it follows that $M = \overline{B_r(0)} \cap g^{-1}(p)$ is compact, $M \subset R^n \setminus f(\partial\Omega) = \cup_{i\geq 1} U_i$ and there exists finitely many i, say $i = 1, 2, \cdots, t$, such that $\cup_{i=1}^{t+1} U_i \supseteq M$, where $U_{t+1} = U_\infty \cap B_{r+1}$. We have

$$deg(f, \Omega, U_{t+1}) = 0, \quad deg(g, U_i, p) = 0$$

for $i \geq t+2$ since $U_j \subset B_r(0)$ and $g^{-1}(y) \cap U_j = \emptyset$ for $j \geq t+2$. Therefore, the right side of (1.2.1) has only finitely many terms different from zero.

We first suppose that $f \in C^1(\overline{\Omega}), g \in C^1(R^n)$ and p is a regular value of gf, so we have

$$deg(gf, \Omega, p) = \sum_{x \in (gf)^{-1}(p)} sgn J_{gf}(x) = \sum_{x \in (gf)^{-1}(p)} sgn J_g(f(x)) sgn J_f(x)$$

and note

$$\begin{aligned}
&\sum_{x \in f^{-1}(z), z \in g^{-1}(p)} sgn J_g(z) sgn J_f(x) \\
&= \sum_{z \in g^{-1}(p), z \in f(\Omega)} sgn J_g(z) [\sum_{x \in f^{-1}(z)} sgn J_f(x)] \\
&= \sum_{z \in f(\Omega), g(z) = p} sgn J_g(z) deg(f, \Omega, z) \\
&= \sum_{i=1}^{t} \sum_{z \in U_i} sgn J_g(z) deg(f, \Omega, z) \\
&= \sum_i deg(f, \Omega, U_i) deg(g, U_i, p).
\end{aligned}$$

For the general case $f \in C(\overline{\Omega})$ and $g \in C(R^n)$, Put

$$V_m = \{z \in B_{r+1}(0) \setminus f(\partial\Omega) : deg(f, \Omega, z) = m\},$$

$$N_m = \{i \in N : deg(f, \Omega, U_i) = m\}.$$

Obviously, $V_m = \cup_{i \in N_m} U_i$ and thus we have

$$\sum_i deg(f, \Omega, U_i) deg(g, U_i, p) = \sum_m [\sum_{i \in N_m} deg(g, U_i, p)] = \sum_m deg(g, V_m, p).$$

Therefore, we need to show

$$deg(gf, \Omega, p) = \sum_m deg(g, V_m, p) \tag{1.2.2}$$

Since $\partial V_m \subset f(\partial\Omega)$, we take $g_0 \in C^1(R^n)$ such that

$$deg(g_0 f, \Omega, p) = deg(gf, \Omega, p), \quad deg(g_0, V_m, p) = deg(g, V_m, p).$$

We may assume that $M_0 = \overline{B_{r+1}(0)} \cap g_0^{-1}(p)$ is not empty and then we have

$$\rho(M_0, f(\partial\Omega)) = \inf\{|x - z| : x \in M_0,\ z \in f(\partial\Omega)\} > 0.$$

Now we take $f_0 \in C^1(\overline{\Omega})$ such that

$$\max_{x \in \overline{\Omega}} |f(x) - f_0(x)| < \rho(M_0, f(\partial\Omega)), \quad f_0(\overline{\Omega}) \subset \overline{B_{r+1}(0)}$$

and put

$$V'_m = \{z \in B_{r+1}(0) \setminus f_0(\partial\Omega) : deg(f_0, \Omega, z) = m\}.$$

Then we have $V_m \cap M_0 = V'_m \cap M_0$ and

$$deg(g_0, V_m, p) = deg(g_0, V_m \cap V'_m, p) = deg(g_0, V'_m, p).$$

Therefore, we have

$$deg(g_0 f_0, \Omega, p) = \sum_m m deg(g_0, V'_m, p) = \sum_m m deg(g, V_m, p).$$

By a simple homotopy argument, one gets $deg(g_0 f_0, \Omega, p) = deg(g_0 f, \Omega, p)$. Thus the conclusion of Theorem 1.2.13 is true. This completes the proof.

By using the product formula, we can prove the following version of Jordan's separation theorem:

Theorem 1.2.14. Let $\Omega_i \subset R^n$, $i = 1, 2$, be two compact sets which are homeomorphic to each other. Then $R^n \setminus \Omega_1$ and $R^n \setminus \Omega_2$ have the same number of connected components.

Proof. Let $h : \Omega_1 \to \Omega_2$ be a homeomorphism onto Ω_2, h' be a continuous extension of h to R^n and $\overline{h^{-1}}$ be a continuous extension of $h^{-1} : \Omega_2 \to \Omega_1$ to R^n. Assume that U_i are bounded components of $R^n \setminus \Omega_1$ and V_j are bounded components of $R^n \setminus \Omega_2$. Notice that $\partial U_i \subset \Omega_1$ and $V_j \subset \Omega_2$. For any fixed i, let W_k denote the components of $R^n \setminus h(\partial U_i)$. Since

$$\cup_j V_j = R^n \setminus \Omega_2 \subset R^n \setminus h(\partial U_i)) = \cup_k W_k$$

for each j, there exists a k such that $V_j \subset W_k$, so, in particular, $V_\infty \subset W_\infty$. For any $p \in U_i$, notice that $\overline{h^{-1}} h' x = x$ for $x \in \partial U_i$, so, by Theorem 1.2.13, we have

$$1 = deg(\overline{h^{-1}} h', U_i, p) = \sum_k deg(h', U_i, W_k) deg(\overline{h^{-1}}, W_k, p).$$

Put $N_k = \{j : V_j \subset W_k\}$. Then we have

$$deg(\overline{h^{-1}}, W_k, p) = \sum_{j \in N_k} deg(\overline{h^{-1}}, V_j, p),$$

$$deg(h', U_i, W_k) = deg(h', U_i, V_j)$$

for all $j \in N_k$. Therefore, we have

$$\begin{aligned} 1 &= \sum_k \sum_{j \in N_k} deg(h', U_i, V_j) deg(\overline{h^{-1}}, V_j, p) \\ &= \sum_j deg(h', U_i, V_j) deg(\overline{h^{-1}}, V_j, U_i), \end{aligned} \tag{1.2.3}$$

since $p \in U_i \subset R^n \setminus (h^{-1}(\Omega)) \subset R^n \setminus h^{-1}(\partial V_j)$. For any fixed j, the same argument implies that

$$1 = \sum_i deg(h', U_i, V_j) deg(h', U_i, V_j) deg(\overline{h^{-1}}, V_j, U_i). \tag{1.2.4}$$

From (1.2.3) and (1.2.4), it follows that $R^n \setminus \Omega_1$ and $R^n \setminus \Omega_2$ have the same number of connected components. This completes the proof.

Theorem 1.2.15. Let $\phi : R^n \to R^1$ be continuously differentiable, $grad\phi(x) \neq 0$ for $|x|$ sufficiently large and $\lim_{|x| \to \infty} \phi(x) = +\infty$. Then

$$\lim_{r \to \infty} deg(grad\phi, B_r(0), 0) = 1.$$

Proof. We may assume that $\phi \in C^\infty(R^n)$. Otherwise, we use the same technique as in Proposition 1.1.5. Define

$$m(x) = \begin{cases} ce^{-\frac{1}{1-|x|}}, & |x| < 1, \\ 0, & |x| \geq 1, \end{cases}$$

where c satisfies $\int_{R^n} m(x)dx = 1$. Set $m_\lambda(x) = \lambda^{-n} m(\frac{x}{\lambda})$ for all $x \in R^n$ and

$$\phi_\lambda(x) = \int_{R^n} \phi(y) m_\lambda(y - x) dx \ \text{ for all } \ x \in R^n, \ \lambda > 0.$$

Then we consider ϕ_λ instead of ϕ for sufficiently small $\lambda > 0$. The initial value problem

$$\begin{cases} u'(t) = -grad\phi(u(t)), \ t > 0, \\ u(0) = x \in R^n \end{cases} \tag{E 1.2.1}$$

has a unique local solution $u(t, x)$ for all $x \in R^n$. Since $\psi(t) = \phi(u(t, x))$ satisfies $\psi'(t) = -|grad\phi(u(t, x))|^2 \leq 0$, we have $\phi(u(t, x)) \leq \phi(x)$ on the interval, where $u(t, x)$ exists. By assumption $\lim_{|x| \to \infty} \phi(x) = +\infty$, $u(t, x)$

can be extended to a unique solution on $[0, \infty)$. We may also assume that $\phi(x) \geq 0$ by adding a constant. Take $r_0 > 0$ such that $grad\phi(x) \neq 0$ for $|x| \geq r_0$, put $M = \max_{x \in \overline{B_{r_0}(0)}} \phi(x)$. Choose $r_1 > r_0$ such that $\phi(x) \geq M + 1$ for $|x| \geq r_1$ and set

$$M_1 = \max_{x \in \partial B_{r_1}(0)} \phi(x).$$

Again, by $\phi(u(t,x)) = \phi(x) - \int_0^t |grad\phi(u(s,x))|^2 ds$, we know that if $x \in \partial B_{r_1}(0)$, then $\phi(u(t,x)) \leq M_1$. Let $\beta = \min\{|grad\phi(x)| : |x| \geq r_0$ and $\varphi(x) \leq M_1\}$. Then, if $|x| = r_1$, we have

$$\phi(u(t,x)) \leq M_1 - \beta^2 t \quad \text{as long as} \quad |u(t,x))| \geq r_0.$$

Thus $|u(t,x)| \leq r_0$ for some $t < \frac{M_1}{\beta^2}$, which implies that

$$u(M_1\beta^{-2}, x) \in B_{r_1}(0) \qquad \text{for all} \quad x \in \partial B_{r_1}(0).$$

This implies that the Poincaré operator $Px = u(M_1\beta^{-2}, x)$ must satisfy $deg(I - P, B_{r_1}(0), 0) = 1$. Next, define

$$h(t,x) = \begin{cases} (x - u(M_1\beta^{-2}t, x)[t + (1-t)(tM_1\beta^{-2})^{-1}], & t \neq 0, \\ grad\phi(x) & t = 0. \end{cases}$$

It is easy to check that h is continuous and $h(t,x) \neq 0$ for all $x \in \partial B_{r_1}(0)$. Thus we have $deg(grad\phi, B_{r_1}(0), 0) = deg(I - P, B_{r_1}(0), 0) = 1$. This completes the proof.

Theorem 1.2.16. Let $\Omega \subset R^n$ be an open subset and $f : \Omega \to R^n$ be continuous and locally one to one. Then f is an open mapping.

Proof. For each $x_0 \in \Omega$, we prove that there exists $r > 0$ such that $f(B_r(x_0))$ contains a ball with center at $f(x_0)$. Without loss of generality, we may assume that $x_0 = 0$. Otherwise, put $\Omega_1 = \Omega - \{x_0\}$ and $f_1(x) = f(x + x_0) - f(x_0)$.

Choose $r > 0$ such that f is one to one on $B_r(0)$. Set $h(t,x) = f(\frac{1}{1+t}x) - f(-\frac{t}{1+t}x)$ for all $(t,x) \in [0,1] \times \overline{B_r(0)}$. Then h is continuous and $h(t,x) \neq 0$ for all $(t,x) \in [0,1] \times \partial B_r(0)$. Otherwise, $h(t,x) = 0$ for some $(t,x) \in [0,1] \times \partial B_r(0)$ and so then $\frac{1}{1+t}x = -\frac{t}{1+t}x$ since f is one to one and $x = 0$, which is a contradiction. Therefore, we have

$$deg(f, B_r(0), 0) = deg(h(1, \cdot), B_r(0), 0) \neq 0$$

since $h(1, \cdot)$ is odd. Choose $t > 0$ such that $t < \inf\{|f(x)| : x \in \partial B_r(0)\}$. Then

$$deg(f, B_r(0), y) = deg(f, B_r(0), 0).$$

Now, we have $B_t \subset f(B_r(0))$ and thus f is open. This completes the proof.

Theorem 1.2.17. Let $0 \in \Omega$ be open bounded and symmetric. If $A_i \subset \partial\Omega$ is closed, $A_i \cap (-A_i) = \emptyset$ for $i = 1, 2, \cdots, k$ and $\cup_{i=1}^k A_i = \partial\Omega$, then $k \geq n+1$.

Proof. Assume the contrary, $k \leq n$. Set $f_i(x) = 1$ on A_i, $f_i(x) = -1$ on $-A_i$ for $i = 1, 2 \cdots, k-1$, $f_i(x) = 1$ on $\overline{\Omega}$ for $i = k, \cdots, n$, and $f = (f_1, f_2, \cdots, f_n)$. Extend f continuously to $\overline{\Omega}$. Then $f(-x) \neq \lambda f(x)$ on $\partial\Omega$ for all $\lambda \geq 0$. Otherwise, $f(-x_0) = \lambda f(x_0)$ for some $\lambda \geq 0$ and $x_0 \in \partial\Omega$. Now, $\lambda > 0$ since $f(x) \neq 0$ on $\partial\Omega$. Also $x_0 \notin A_i \cup (-A_i)$ for $i \leq k-1$ since $f_i(-x) = -f_i(x)$. Thus $x_0 \in A_k$. Also $x_0 \notin -A_k$, so we have $-x_0 \in A_i$ for some $i \leq k-1$ and thus $x_0 \in -A_i$, which is a contradiction. Therefore, $f(-x) \neq \lambda f(x)$ on $\partial\Omega$ for all $\lambda \geq 0$. Thus we have $deg(f, \Omega, 0) = 0$, i.e., $f(x) = 0$ for some $x \in \Omega$, which is a contradiction to $f_n(x) = 1$ on $\overline{\Omega}$. This completes the proof.

Next, we prove that the winding number is a special case of the Brouwer degree.

Theorem 1.2.18. Let $B(0,1) \subset C$ be the unit ball, $\Gamma = \partial B(0,1)$ and $f : \overline{B(0,1)} \to C$ be a C^1 function. Assume that $a \notin f(\Gamma)$. Then

$$deg(f, B(0,1), a) = \frac{1}{2\pi i} \int_{f(\Gamma)} \frac{1}{z-a} dz. \tag{1.2.5}$$

Proof. It is sufficient to prove (1.2.5) in the case when $a \notin f(S_f)$. Let $f^{-1}(a) = \{z_1, z_2, \cdots, z_k\}$. Then we need to show

$$\frac{1}{2\pi i} \int_{f(\Gamma)} \frac{1}{z-a} dz = \sum_{i=1}^{k} sgn J_f(z_i). \tag{1.2.6}$$

Take $\epsilon > 0$ small enough such that the $\overline{V_i}$'s are disjoint, where $V_i = B(z_i, \epsilon)$, $sgn J_f(z) = sgn J_f(z_i)$ for $z \in V_i$, $\overline{V_i} \subset B(0,1)$ and the restriction of f to $\overline{V_i}$ is a homeomorphism for $i = 1, 2, \cdots, k$. Put $S_i = \partial V_i$. Then $f(S_i)$ is a Jordan curve such that a lies in its interior region, $f(S_i)$ has the same orientation as S_i if $J_f(z_i) > 0$ and the opposite orientation if $J_f(z_i) < 0$.

Now, set $U = \overline{B(0,1)} \setminus \cup_{i=1}^k V_i$. Then $|f(z) - a| > \alpha$ in U for some $\alpha > 0$. We can divide U into small rectangles R_j such that $|f(z) - f(w)| < \alpha$ on each R_j. Since the image $f(\partial(R_j \cap V))$ does not wind around a, we have $w(f(\partial(R_j \cap V)), a) = 0$, and summing over all R_j yields

$$\int_{f(\Gamma)} \frac{1}{z-a} dz = \sum_{i=1}^{k} \int_{f(S_i)} \frac{1}{z-a} dz.$$

Since the orientation of $f(S_i)$ is determined by $J_f(z_i)$, $f(S_i)$ winds exactly once around a. Thus, we have

$$\int_{f(S_i)} \frac{1}{z-a} dz = sgn J_f(z_i)$$

and so (1.2.5) is true. This completes the proof.

Remark. (1.2.5) is also correct if f is continuous on $\overline{B(0,1)}$.

Theorem 1.2.19. Let $B(0,1) \subset C$ be the unit ball, $\Gamma = \partial B(0,1)$ and $f(z) = \sum_{n=0} a_n z_n : \overline{B(0,1)} \to C$ with $\sum_{n=1}^{\infty} n a_n \overline{a_n} < \infty$. Suppose that $f(\Gamma) \subset \Gamma$. Then $\sum_{n=1}^{\infty} n a_n \overline{a_n}$ is a non-negative integer.

Proof. Since $0 \notin \Gamma$, by Theorem 1.2.18, we have

$$deg(f, B(0,1), 0) = \frac{1}{2\pi i} \int_{f(\Gamma)} \frac{1}{z} dz$$

and

$$\frac{1}{2\pi i} \int_{f(\Gamma)} \frac{1}{z} dz = \frac{1}{2\pi i} \int_{\Gamma} \frac{f'(z)}{f(z)} dz = \frac{1}{2\pi i} \int_{\Gamma} f'(z)\overline{f(z)} dz = \sum_{n=1}^{\infty} n a_n \overline{a_n}.$$

Thus the conclusion is true. This completes the proof.

1.3 Degree Theory for Functions in VMO

Let $\Omega \subset R^n$ be an open bounded subset and $f : \overline{\Omega} \to R^n$ a measurable function such that $\int_{\Omega} |f(x)| dx < \infty$. For any ball $B_r(x) \subset \Omega$, we define $A_r f(x)$ to be the average value of f as follows:

$$A_r f(x) = \frac{1}{m(B_r(x))} \int_{B_r(x)} f(y) dy.$$

Lemma 1.3.1. $A_r f(x)$ is continuous in r for each x and measurable in x for each r.

Proof. Since $m(B_r(x)) = r^n m(B(0,1))$ and $m(\partial B_r(x)) = 0$, we have $\chi_{B_r(x)}(y) \to \chi_{B_s(x)}(y)$ almost everywhere as $r \to s$, where $\chi_E(y) = 1$ if $y \in E$, while $\chi_E(y) = 0$ if $y \notin E$. By Lebesgue's dominated convergence theorem, we know that $A_r f(x)$ is continuous in r. Also, we have

$$A_r f(x) = r^{-n} (m(B(0,1)))^{-1} \int_{\Omega} \chi_{B_r(x)(y)} f(y) dy,$$

$\chi_{B_r(x)(y)}$ is clearly measurable, so the measurability of $A_r f(x)$ follows from Fubini's theorem. This completes the proof.

Lemma 1.3.2. Let ϕ be any collection of open balls in R^n and $U = \cup_{B \in \phi} B$. If $0 < c < m(U)$, then there exist disjoint $B_1, B_2, \cdots, B_k \in \phi$ such that $\sum_{i=1}^{k} m(B_i) > 3^{-n} c$.

Proof. Since $c < m(U)$, there exists a compact subset $K \subset U$ with $m(K) > c$, and finitely many of the balls in ϕ, say, $A_1, A_2, \cdots, A_m$, cover K. Let B_1 be the largest of the A_i's (that is, choose B_1 to have maximal radius), let B_2 be the largest of the A_i's which are disjoint from B_1 and so on until the sequence of A_i's is exhausted. If A_j is not one of the B_i's, there exists i such that $A_j \cap B_i \neq \emptyset$ and the radius of A_j is at most that of B_i. Therefore, $A_j \subset B_i^*$, where B_i^* is the ball concentric with B_i whose radius is three times that of B_i. But then $K \subset \cup_i B_i^*$, so

$$c < m(K) \leq \sum_i m(B_i^*) = 3^n \sum_i m(B_i).$$

This completes the proof.

Next, if $f \in L^1(\Omega)$, we define its Hardy Littlewood maximal function Hf by

$$Hf(x) = \sup_{r>0} A_r|f|(x) = \sup_{r>0} \frac{1}{m(B_r(x))} \int_{B_r(x)} |f(y)|dy.$$

Theorem 1.3.3. There is a constant $\beta > 0$ such that, for all $f \in L^1(\Omega)$ and $\alpha > 0$,

$$m(\{x : Hf(x) > \alpha\}) \leq \frac{\beta}{\alpha} \int_\Omega |f(x)|dx.$$

Proof. Let $E_\alpha = \{x : Hf(x) > \alpha\}$. For each $x \in E_\alpha$, we can choose $r_x > 0$ such that $A_{r_x}|f|(x) > \alpha$. The balls $B_{r_x}(x)$ cover E_α and so, by Lemma 1.3.2, if $c < m(E_\alpha)$, there exist $x_1, x_2, \cdots, x_k \in E_\alpha$ such that the balls $B_i = B_{r_{x_i}}(x_i)$ are disjoint and $\sum_{i=1}^k m(B_i) > 3^{-n}c$. But then

$$c < 3^n \sum_{i=1}^k m(B_i) \leq \frac{3^n}{\alpha} \sum_{i=1}^k \int_{B_i} |f(y)|dy \leq \frac{3^n}{\alpha} \int_\Omega |f(y)|dy.$$

By letting $c \to m(E_\alpha)$, we obtain the desired result. This completes the proof.

Theorem 1.3.4. If $f \in L^1(\Omega)$, then $\lim_{r\to 0} A_r f(x) = f(x)$ for almost all $x \in \Omega$.

Proof. For any $\epsilon > 0$, there exists a continuous function g such that

$$\int_\Omega |f(x) - g(x)|dx < \epsilon.$$

Continuity of g implies that for every $x \in \Omega$ and $\delta > 0$, there exists $r > 0$ such that $|g(y) - g(x)| < \delta$ whenever $|y - x| < r$, and hence

$$|A_r g(x) - g(x)| < \delta.$$

Therefore, $A_r g(x) \to g(x)$ as $r \to 0$ for all $x \in \Omega$, so we have

$$\begin{aligned}&\limsup_{r\to 0} |A_r f(x) - f(x)| \\ &= \limsup_{r\to 0} |A_r(f-g)(x) + (A_r g(x) - g(x)) + (g-f)(x)| \\ &\le H(f-g)(x) + |f(x) - g(x)|.\end{aligned}$$

Hence, if

$$E_\alpha = \{x : \limsup_{r\to 0} |A_r(f-g)(x) > \alpha\}, \quad F_\alpha = \{x : |f(x) - g(x)| > \alpha\},$$

then we have

$$E_\alpha \subset F_{\frac{\alpha}{2}} \cup \{x : H(f-g)(x) > \frac{\alpha}{2}\}.$$

However, $\alpha m(F_\alpha) \le \int_{F_\alpha} |f(x) - g(x)|dx < \epsilon$ and so, by Theorem 1.3.3, we have

$$m(E_\alpha) \le \frac{2\epsilon}{\alpha} + \frac{2\beta\epsilon}{\alpha}.$$

By letting $\epsilon \to 0$, we get $m(E_\alpha) = 0$ for all $\alpha > 0$. This completes the proof.

In the following, let Ω be a smooth open bounded domain in R^n or a smooth compact Riemannian manifold. (For the definition of Riemannian manifold, we refer the reader to [171].

Next, we introduce the following concept:

Definition 1.3.5. Let $f \in L^1(\Omega)$ and $B_r \subset \Omega$ a (geodesic) ball with radius $r > 0$. If $\sup_{r>0} \frac{1}{m(B_r)} \int_{B_r} \frac{1}{m(B_r)} \int_{B_r} |f(x) - f(y)|dxdy < \infty$, then f is called a bounded mean oscillation function. The set of all bounded mean oscillation functions is denoted by BMO.

If $\lim_{r\to 0} \frac{1}{m(B_r)} \int_{B_r} \frac{1}{m(B_r)} \int_{B_r} |f(x) - f(y)|dxdy = 0$, then f is called a vanishing mean oscillation function. The set of all vanishing mean oscillation functions is denoted by VMO.

Example 1.3.6. If $f \in L^1(\Omega)$, then $f \in BMO$.

Example 1.3.7. $f(x) = |\log|x|| \in BMO$.

In the following, $1 \le p < +\infty$, let $W^{1,p}(\Omega) = \{u(\cdot) : \Omega \to R$ such that $u(\cdot) \in L^p(\Omega), u'(\cdot) \in L^p(\Omega)\}$.

Proposition 1.3.8. If $f(\cdot) \in W^{1,n}$, then $f(\cdot) \in VMO$.

Proof. By Poincaré's inequality, we have

$$\int_{B_r} |f(y) - \frac{1}{m(B_r)} \int_{B_r} |f(x)|dx|dy \le cm(B)^{\frac{1}{n}} \int_{B_r} |\nabla f(x)|dx.$$

Then we deduce, by using the Hölder inequality, that

$$\frac{1}{m(B_r)}\int_{B_r}|f(y)-\frac{1}{m(B_r)}\int_{B_r}|f(x)|\leq c(\int_\Omega|\nabla f(x)|dx)^{\frac{1}{n}}.$$

Thus the conclusion is true.

Recall that, for $0<s<1$, $1<p<+\infty$, the fractional Sobolev space $W^{s,p}(\Omega)$ is characterized by

$$W^{s,p}(\Omega)=\{f(\cdot)\in L^p(\Omega),\int_\Omega\int_\Omega\frac{|f(x)-f(y)|^p}{|x-y|^{n+sp}}<+\infty\}.$$

Proposition 1.3.9. If $sp=n$, then $W^{s,p}(\Omega)\subset VMO$.

Proof. Clearly, one has

$$\begin{aligned}&\int_B\int_B|f(x)-f(y)|dxdy\\&=\int_B\int_B\frac{|f(x)-f(y)|}{|x-y|^{(\frac{n}{p})+s}}|x-y|^{(\frac{n}{p})+s}dxdy\\&\leq c(m(B))^{\frac{1}{p}+\frac{s}{n}}\int_B\int_B\frac{|f(x)-f(y)|}{|x-y|^{(\frac{n}{p})+s}}dxdy\end{aligned}$$

for some constant $c>0$. By using Hölder's inequality, we deduce that

$$\int_B\int_B\frac{|f(x)-f(y)|}{|x-y|^{(\frac{n}{p})+s}}dxdy\leq cm(B)^{\frac{1}{p}+\frac{s}{n}+2-\frac{2}{p}}\int_B\int_B\frac{|f(x)-f(y)|^p}{|x-y|^{n+sp}}$$

and thus, when $sp=n$,

$$\frac{1}{m(B)^2}\int_B\int_B|f(x)-f(y)|dxdy\leq c[\int_B\int_B\frac{|f(x)-f(y)|^p}{|x-y|^{n+sp}}]^{\frac{1}{p}}.$$

This completes the proof.

Lemma 1.3.10. Let $B(0,1)$ be the unit ball of R^{n+1}, $S^n=\partial B(0,1)$ and $f\in L^1(S^n,S^n)$ be such that $f\in VMO$. Then $A_rf(x)=\frac{1}{m(B_r(x))}\int_{B_r(x)}f(y)dy$ is continuous in x for small $r>0$ and $\lim_{r\to0}|A_rf(x)|=1$ uniformly on S^n.

Proof. The continuity of $A_rf(x)$ is guaranteed since $f\in VMO$. To prove the uniform convergence of $A_rf(x)$ on S^n, we set

$$\delta_r(x)=\frac{1}{m(B_r(x))^2}\int_{B_r(x)}\int_{B_r(x)}|f(y)-f(z)|dydz.$$

Then $\delta_r(x)\to0$ as $r\to0$ uniformly on S^n and we have

$$1-\delta_r(x)\leq|A_rf(x)|\leq1.$$

Thus $|A_r f(x)| \to 1$ uniformly on S^n as $r \to 0$. This completes the proof.

Now, assume that $f \in L^1(S^n, S^n)$ and $f \in VMO$. By Lemma 1.3.10, there exists $r_0 > 0$ such that $A_r f(x) \neq 0$ for all $x \in S^n$ and $0 < r < r_0$. Let $\tilde{A}_r f(x)$ be a continuous extension of $A_r f(x)$ to $\overline{B(0,1)}$. Then the Brouwer $deg(\tilde{A}_r f, B(0,1), 0)$ is well defined for $r \in (0, r_0)$. It does not depend on the extension $\tilde{A}_r f$. Consider the homotopy $\{\tilde{A}_{tr_1+(1-t)r_2} f\}_{t\in[0,1]}$ for all $r_1, r_2 \in (0, r_0)$. It follows from Lemma 1.3.10 that $A_{tr_1+(1-t)r_2} f(x) \neq 0$ for all $(t,x) \in [0,1] \times S^n$ and thus $deg(\tilde{A}_r f, B(0,1), 0)$ does not depend on $r \in (0, r_0)$. Now, we define the topological degree by

$$deg(f, S^n) = \lim_{r\to 0} deg(\tilde{A}_r f, B(0,1), 0). \tag{1.3.1}$$

Proposition 1.3.11. We have

$$\lim_{r\to 0} deg(\tilde{A}_r f, B(0,1), 0) = \lim_{r\to 0} deg(\frac{\tilde{A}_r f}{|\tilde{A}_r f|}, B(0,1), 0).$$

Proof. Consider the homotopy $H_r(t,x) = \frac{\tilde{A}_r f(x)}{(1-t)+t|\tilde{A}_r f(x)|}$ for all $(t,x) \in [0,1] \times \overline{B(0,1)}$. By Lemma 1.3.10, we know that $H_r(t,x) \neq 0$ for all $(t,x) \in [0,1] \times S^n$ for r sufficiently small and so the conclusion follows from Theorem 1.2.6.

Remark. For more results regarding the degree defined by (1.3.1), we refer the reader to Brezis and Nirenberg [29] (see also [25], [27], [28], [175] for more results on the computation of the degree for Sobolev maps).

1.4 Applications to ODEs

In this section, we give some applications of results in section 1.2 to periodic and anti-periodic problems of ordinary differential equations in R^n.

Theorem 1.4.1. Let $f : R \times R^n \to R^n$ be a continuous function and $f(t+T, x) = f(t,x)$ for all $(t,x) \in R \times R^n$. Suppose that the following conditions are satisfied:

(1) There exists $r > 0$ such that $(f(t,x), x) < 0$ for all $t \in [0,T]$ and $|x| = r$.

(2) For each $x \in R^n$, there exist $r_x > 0$, $L_x > 0$ such that

$$|f(t,y) - f(t,z)| \le L_x|y - z| \quad \text{for all } t \in R,\ y, z \in B(x, r_x).$$

Then the following equation:

$$\begin{cases} x'(t) = f(t, x(t), \ t \in [0, +\infty), \\ x(0) = x(T) \end{cases} \tag{E 1.4.1}$$

has a solution.

Proof. For each $x_0 \in \overline{B(0,r)}$, by Peano's Theorem, the initial value problem

$$\begin{cases} x'(t) = f(t, x(t)), \ \ t \in (0, t_0), \\ x(0) = x_0 \end{cases} \tag{E 1.4.2}$$

has a solution for some $t_0 > 0$. If (E 1.4.2) has two solutions $x(\cdot), y(\cdot)$, then

$$\frac{d}{dt}|x(t) - y(t)|^2 = 2(x'(t) - y'(t), x(t) - y(t)) \leq L_{x_0}|x(t) - y(t)| \tag{1.4.1}$$

for some $t_1 \in (0, t_0)$ and $t \in (0, t_1)$. From (1.4.1), we get

$$|x(t) - y(t)| \leq e^{L_{x_0} t}|x(0) - y(0)| \quad \text{for all } t \in (0, t_1),$$

so $x(t) = y(t)$ for all $t \in (0, t_1)$. Therefore, $x(t) = y(t)$ for $t \in [0, t_0]$, so the solution of (E 1.4.2) is unique.

If $x(t) = r$, then $\frac{d}{dt}|x(t)|^2 = 2(x'(t), x(t)) = (f(t, x(t)), x(t)) < 0$. Thus $x(t)$ must stay in $\overline{B(0,r)}$ for all $t \in [0, t_0]$, so $x(t)$ can be extended to $[0, +\infty)$ and also $x(t) \in \overline{B(0,r)}$ for $t \in [0, +\infty)$.

Now, we define a mapping $S : \overline{B(0,r)} \to \overline{B(0,r)}$ as follows:

$$Sy = x(y, T) \qquad \text{for all } \ y \in \overline{B(0,r)},$$

where $x(y,t)$ is the unique solution of (E 1.4.2) with initial value y.

Again, by using (1.4.1), one can easily prove that S is continuous. Thus, by Brouwer's fixed point theorem, S has a fixed point in $\overline{B(0,r)}$, i.e., (E 1.4.1) has a solution. This completes the proof.

Theorem 1.4.2. Let $G : R^n \to R$ be an even continuous differentiable function such that ∂G is Lipschitzian and $f : R \to R^n$ be a continuous function such that $f(t+T) = -f(t)$ for all $t \in R$, then the following equation:

$$\begin{cases} x'(t) = \partial G x(t) + f(t), & t \in R, \\ x(t+T) = -x(t), & t \in R \end{cases} \tag{E1.4.3}$$

has a solution.

Proof. First, if $x(t)$ is a solution of (E 1.4.3), then

$$|x'(t)|^2 = (\partial G x(t), x'(t)) + (f(t), x'(t)) \quad \text{for all } t \in R.$$

Integrate over $[0,T]$ and notice that $\int_0^T (\partial Gx(t), x'(t))dt = 0$, we have

$$\int_0^T |x'(t)|^2 dt = \int_0^T (f(t), x'(t))dt$$

and thus $(\int_0^T |x'(t)|^2 dt)^{\frac{1}{2}} \leq (\int_0^T |f(t)|^2 dt)^{\frac{1}{2}}$.

In addition, $2x(t) = \int_0^t x'(s)ds - \int_T^t x'(s)ds$ and thus we have

$$\max_{t\in[0,T]} |x(t)| \leq \frac{\sqrt{T}}{2}(\int_0^T |f(t)|^2 dt)^{\frac{1}{2}} = M.$$

We take an even continuous differentiable function ϕ with $\phi(x) = 1$ for $|x| \leq M$ and $\partial\phi(x) = 0$ for $|x| > 2M$.

Now, we consider the following equation:

$$\begin{cases} x'(t) + x(t) = \partial(\phi(x(t))[G(x(t)) + \frac{1}{2}|x(t)|^2]) + f(t), \ \ t \in R, \\ x(0) = y \in R^n. \end{cases} \qquad (E\ 1.4.4)$$

Since $\partial\phi$ is uniformly bounded, we have, for each solution $x(t)$ of $(E\ 1.4.4)$,

$$\frac{d}{dt}(|x(t)|^2) + |x(t)|^2 \leq 2L|x(t)| + 2|f(t)||x(t)|$$

for some constant $L > 0$. Thus, if there exists a $N > 0$ such that $|x(0)| \leq N$, then $|x(T)| \leq N$.

Now, we define a map $S : R^n \to R^n$ by $Sy = -x(T)$, where $x(\cdot)$ is the unique solution of $(E\,1.4.4)$ with $x(0) = y$. It is obvious that S is continuous, so, by Brouwer's fixed point theorem, Theorem 1.2.7, there exists $y \in R^n$ such that $Sy = y$, i.e., $x(T) = -y$. Thus $|x(t)| \leq M$ for all $t \in R$ and, consequently, it follows that

$$\partial(\phi(x(t))[G(x(t)) + \frac{1}{2}|x(t)|^2]) = \partial Gx(t) + x(t).$$

Therefore, $x(\cdot)$ is a solution of $(E\,1.4.3)$. This completes the proof.

Corollary 1.4.3. Let A be a $n \times n$ symmetric matrix and $f : R \to R^n$ be a continuous function such that $f(t+T) = -f(t)$ for all $t \in R$. Then the following equation:

$$\begin{cases} x'(t) = Ax(t) + f(t), & t \in R, \\ x(t+T) = -x(t), & t \in R \end{cases} \qquad (E\ 1.4.5)$$

has a solution.

Proof. Since A is symmetric, put $Gu = \frac{1}{2}(Au, u)$ for $u \in R^n$ and then $A = \partial G$. Thus the conclusion follows from Theorem 1.4.2.

Example 1.4.4. The following equation:

$$\begin{cases} x_1'(t) = 2x_1(t) - \alpha x_2(t) + sint, & t \in R, \\ x_2'(t) = -\alpha x_1(t) - 3x_2(t) + sin^3 t, & t \in R, \\ x_1(t+\pi) = -x_1(t), x_2(t+\pi) = -x_2(t), & t \in R, \end{cases} \qquad (E\ 1.4.6)$$

has a solution, where $\alpha \in R$ is a constant. In fact, set

$$A = \begin{pmatrix} 2 & -\alpha \\ -\alpha & -3 \end{pmatrix}, \quad x = \begin{pmatrix} x_1 \\ x_2 \end{pmatrix}, \quad f(t) = \begin{pmatrix} \sin t \\ \sin^3 t \end{pmatrix}.$$

Then A is symmetric and the conclusion follows from Corollary 1.4.3.

1.5 Exercises

1. Let $\Omega \subset R^n$ be open bounded and $0 \in \Omega$ and $f : \overline{\Omega} \to R^n$ be continuous and $(f(x), x) \geq 0$. Show $0 \in f(\overline{\Omega})$.

2. Let $\Omega \subset R^2$ be open bounded and $u(x,y), v(x,y) : \overline{\Omega} \to R$ be continuously differentiable functions with $u_x = v_y$ and $u_y = -v_x$. Assume that $f(x,y) = (u(x,y), v(x,y)) : \overline{\Omega} \to R^2$ has m many zero points in $\Omega \setminus S_f(\Omega)$. Show that $deg(f, \Omega, 0) = m$.

3. Prove the fundamental theorem of algebra by using Brouwer degree.

4. Let $B(0,1)$ be the unit ball in R^n and $f : R^n \to R^n$ be a continuous function such that $f(\partial B(0,1)) = \partial B(0,1)$. Show that $deg(f^m, B(0,1), 0) = (deg(f, B(0,1), 0)^m$ for all positive integer m.

5. For any integer n, show that there exists an open bounded subset $\Omega \subset R$ and a continuous function $f : \overline{\Omega} \to R$ such that $deg(f, \Omega, 0) = n$.

6. Let $\Omega \subset R^n$ be open bounded, $f, g \in C(\overline{\Omega})$ and $|g(x)| < |f(x)|$ for all $x \in \partial\Omega$. Show $deg(f - g, \Omega, 0) = deg(f, \Omega, 0)$.

7. Let $B(0,1)$ be the unit ball of R^{2n+1} and $f : \partial B(0,1) \to \partial B(0,1)$ be continuous. Show that there exists $x_0 \in \partial B(0,1)$ such that $f(x_0) = x_0$ or $f(x_0) = -x_0$.

8. Let $\Omega \subset R^n$ be open bounded symmetric, $0 \in \Omega$ and $f : \partial\Omega \to R^m$ be a function with $m < n$. Show that $f(x) = f(-x)$ for some $x \in \partial\Omega$

9. Let $\Omega \subset R^n$ be open bounded and $f \in C(\overline{\Omega})$. Suppose that there exists $x_0 \in \Omega$ such that f satisfies the following condition:

   $$f(x) - x_0 = t(x - x_0) \quad \text{for some } x \in \partial\Omega.$$

   Then $t \leq 1$. Show that f has a fixed point in $\overline{\Omega}$.

10. Let $\phi : \overline{B(0,1)} \subset R^n \to R^n$ be a continuous function such that $\phi(\partial B(0,1)) \subset R^n \setminus \{0\}$, $\psi(x) = \frac{\phi(x)}{|\phi(x)|}$ for all $x \in \partial B(0,1)$ and $\partial B(0,1) \neq \psi(\partial B(0,1))$. Show that
$$deg(\phi, B(0,1), 0) = deg(\psi, B(0,1), p)$$
for all $p \in \partial B(0,1) \setminus \psi(\partial B(0,1))$.

11. If $f \in L^1(\Omega)$, show that $\lim_{r\to 0} \frac{1}{m(B_r(x))} \int_{B_r(x)} |f(y) - f(x)|dy = 0$ for almost all $x \in \Omega$.

12. If $f(\theta) = \sum_{n=0}^{\infty} a_i e^{in\theta} \in L^2(S^1, S^1)$ and $\sum_{n=0}^{\infty} n|a_n|^2 < +\infty$, show that
$$deg(f, S^1) = \sum_{n=0}^{\infty} n|a_n|^2.$$

13. Show that the following equation:
$$\begin{cases} x'(t) = x^3(t) + sin^5 t, & t \in R, \\ x(t+\pi) = -x(t) & t \in R, \end{cases}$$
has a solution.

Chapter 2

LERAY SCHAUDER DEGREE THEORY

Many problems in science lead to the equation $Tx = y$ in infinite dimensional spaces rather than to the finite dimensional case in Chapter 1. In particular, ordinary and partial differential equations, and integral equations can be formulated as abstract equations on infinite dimensional spaces of functions. For the equation $Tx = y$, we again are interested in the questions raised at the beginning of Chapter 1.

In 1934, Leray and Schauder [185] generalized Brouwer degree theory to an infinite Banach space and established the so-called the Leray Schauder degree. It turns out that the Leray Schauder degree is a very powerful tool in proving various existence results for nonlinear partial differential equations (see [135], [185], [203], [228], etc.).

In this chapter, we will introduce the Leray Schauder degree. This chapter consists of five sections.

Section 2.1 gathers together some well known results on compact maps.

In Section 2.2, we first show how a compact map can be approximated by maps with finite dimensional ranges and from here we define the Leray Schauder degree for compact maps. The main properties of this degree are presented in theorems 2.2.4, 2.2.8 and 2.2.16. Also, various consequences, for example, Schauder's fixed point theorem, the Leray Schauder alternatives and compression and expansion fixed point theorems, are presented in this section.

Section 2.3 presents a degree theory for multi-valued maps, and the theory is based on the fact that a upper semicontinuous map admits an approximate continuous selection (see Lemma 2.3.7). We use the degree theory in this chapter to discuss bifurcation problems in Section 2.4 and ordinary (initial and anti-periodic) and partial differential equations in Section 2.5.

2.1 Compact Mappings

In this section, we give some properties of compact operators in topological spaces.

Definition 2.1.1. Let X be a topological space. A subset $M \subset X$ is called compact if every open covering of M has an finite covering, i.e., if $M \subset \cup_{i\in I} V_i$, where V_i is an open subset of X for all $i \in I$, then there exist $i_j \in I$, $j = 1, 2, \cdots, k$, such that $M \subset \cup_{j=1}^k V_{i_j}$.

M is called relatively compact if $\overline{M}$ is compact.

Definition 2.1.2. Let X be a nonempty subset and $d(\cdot, \cdot) : X \times X \to R$ be a function satisfying the following conditions:

(1) $d(x, y) \geq 0$ for all $x, y \in X$ and $d(x, y) = 0$ iff $x = y$;

(2) $d(x, y) = d(y, x)$ for all $x, y \in X$;

(3) $d(x, y) \leq d(x, z) + d(z, y)$ for all $x, y, z \in X$.

Then we call d a *metric* on X and (X, d) a metric space.

Let (X, d) be a metric space, $x \in X$ and $r > 0$. Let $B(x, r) = \{y \in X : d(x, y) < r\}$ be a open ball with center x and radius r.

Proposition 2.1.3. Let (E, d) be a metric space. Then, a subset $M \subset X$ is compact if and only if every infinite sequence $(x_n)_{n=1}^\infty \subset M$ has a convergent subsequence in M.

Proof. Assume that M is compact and $(x_n)_{n=1}^\infty \subset M$. If $(x_n)_{n=1}^\infty$ does not have a convergent subsequence in M, then, for any $y \in M$, there exist $r_y > 0$ and an integer $N_y > 0$ such that $B(y, r_y) \cap \{x_n : n \geq n_y\} = \emptyset$. Notice that $\cup_{y \in M} B(y, r_y) \supset M$, so there exist finitely many $y_1, y_2, \cdots, y_k$ such that $M \subset \cup_{i=1}^k B(y_i, r_{y_i})$. However, $B(y_i, r_{y_i}) \cap \{x_n : n \geq m\} = \emptyset$ for $i = 1, 2, \cdots, k$, where $m = \max\{n_{y_i} : i = 1, 2 \cdots, k\}$. Thus

$$\begin{aligned}
\{x_n : n \geq m\} &= M \cap \{x_n : n \geq m\} \\
&\subset \cup_{i=1}^k B(y_i, r_{y_i}) \cap \{x_n : n \geq m\} \\
&= \emptyset,
\end{aligned}$$

which is a contradiction and so $(x_n)_{n=1}^\infty$ has a convergent subsequence in M.

On the other hand, assume that every infinite sequence $(x_n)_{n=1}^\infty \subset M$ has a convergent subsequence in M. We prove that M is compact. Let $(U_i)_{i\in I}$ be an open covering of M. For any $x \in M$, there exists U_i such that $x \in U_i$. Since U_i is open, there exists $r > 0$ such that $B(x, r) \subset U_i$. Put $r_x = \sup\{r > 0 : B(x, r) \subset B_i$ for some $i \in I\}$ and set $r_0 = \inf\{r_x : x \in M\}$. We next prove that $r_0 > 0$. There exists a sequence $(x_n)_{n=1}^\infty \subset M$ such that $r_i = r_{x_i} \to r_0$. By assumption, there exists a subsequence $(x_{n_k})_{k=1}^\infty$ such that $x_{n_k} \to y_0 \in M$, so there exists an integer $N > 0$ such that $x_{n_k} \in B(x_0, 4^{-1} r_{x_0})$ for $k > N$. Thus it follows that $B(x_{n_k}, 4^{-1} r_{x_0}) \subset B(x_0, 2^{-1} r_{x_0}) \subset B_i$ for some $i \in I$. Consequently, we have $r_0 \geq 2^{-1} r_{x_0}$. For any $x_1 \in M$, if $B(x_1, 2^{-1} r_0) \not\supset M$, there exists $x_2 \in M \setminus B(x_1, 2^{-1} r_0)$, and if $\cup_{i=1}^2 B(2^{-1} x_i, r_0) \not\supset M$, there exists $x_3 \in M \setminus \cup_{i=1}^2 B(2^{-1} x_i, r_0)$, and we claim that this process will terminate at some finite step. If not, there exist $x_n \in M \setminus \cup_{i=1}^{n-1} B(x_i, 2^{-1}r_0)$ for $n \geq 4$, so we have

$$d(x_n, x_m) \geq 2^{-1}r_0 \ for \ n \neq m,$$

which is a contradiction to our assumption. Thus, there exist finitely many $x_1, x_2, \cdots, x_n$ such that $M \subset \cup_{i=1}^{n} B(x_i, 2^{-1}r_0) \subset \cup_{i=1}^{n} U_{j_i}$ and M is compact. This completes the proof.

Definition 2.1.4. Let E be a real vector space. A function $\|\cdot\| : E \to R$ satisfying the following conditions:

(1) $\|x\| \geq 0$ for all $x \in E$, and $\|x\| = 0$ if and only if $x = 0$;

(2) $\|\alpha x\| = |\alpha|\|x\|$ for all $\alpha \in R$, $x \in E$,

(3) $\|x + y\| \leq \|x\| + \|y\|$ for all $x, y \in E$

is called a norm on E and $(E, \|\cdot\|)$ a real normed space or, simply, E is a normed space. If E is also complete, then we say E a real Banach space.

Now, we give some well-known Banach spaces in functional analysis.

Example 2.1.5. Let $1 \leq p < \infty$ and

$$l^p = \{(x_i) : x_i \in R \text{ for } i = 1, 2, \cdots, \ \Sigma_{i=1}^{\infty}|x_i|^p < +\infty\}.$$

Then l^p is a Banach space.

Example 2.1.6. Let

$$c = \{(x_i) : x_i \in R \text{ for } i = 1, 2, \cdots, \text{ and } \lim_{i\to\infty} x_i \text{ exists}\}.$$

Then c is a Banach space.

Example 2.1.7. Let $c_0 = \{(x_i) : x_i \in R$ for $i = 1, 2, \cdots,$ and $\lim_{i\to\infty} x_i = 0\}$. Then c_0 is a Banach space.

Example 2.1.8. Let $\Omega \subset R^n$ be a bounded measurable subset, $1 \leq p < \infty$, and $L^p(\Omega) = \{f(\cdot) : \Omega \to R$ such that $\int_\Omega |f(x)|^p dx < \infty\}$. Then $L^p(\Omega)$ is a Banach space.

Lemma 2.1.9. (Riesz's Theorem) Let E be a real normed space and $M \subset E$ be a proper closed subspace. Then, for any $\epsilon \in (0, 1)$, there exists $x_0 \in E$ such that $\|x_0\| = 1$ and $d(x_0, M) = \inf_{y\in M} \|x_0 - y\| > \epsilon$.

Proof. Since $M \neq E$, there exists $y_0 \in E \setminus M$ such that $d(y_0, M) = \delta_0 > 0$. Take $y_1 \in M$ such that $\|y_0 - y_1\| < \epsilon^{-1}\delta_0$ and put $x_0 = \frac{y_0 - y_1}{\|y_0 - y_1\|}$. Then

$\|x_0\| = 1$ and

$$\begin{aligned} d(x_0, M) &= \inf_{y \in M} \| \frac{y_0 - y_1}{\|y_0 - y_1\|} - y\| \\ &= \inf_{y \in M} \|y_0 - y_1\|^{-1} \|y_0 - y_1 - \|y_0 - y_1\| y\| \\ &> \epsilon \delta_0^{-1} \delta_0 \\ &= \epsilon, \end{aligned}$$

which is the desired result. This completes the proof.

As a consequence of Lemma 2.1.9, we get the following:

Proposition 2.1.10. Let E be a real normed space. Then, the unit closed ball $\overline{B(0,1)} = \{x : \|x\| \leq 1\}$ is compact if and only if $dim(E) < +\infty$.

Definiton 2.1.11. Let E be a real normed space. A mapping $T : D(T) \subset E \to E$ is called compact if T maps every bounded subset of $D(T)$ to a relatively compact subset in E. T is said to be completely continuous if T is continuous and compact.

Definition 2.1.12. Let X, Y be two real Banach spaces and $\Omega \subset X$ be an open subset. A mapping $F : \Omega \to Y$ is said to be Fréchet differentiable at $x_0 \in \Omega$ if there is an $F'(x_0) \in L(X, Y)$ such that

$$F(x_0 + h) = Fx_0 + F'(x_0)h + \delta(x_0, h), \quad \lim_{h \to 0} \frac{\delta(x_0, h)}{\|h\|} = 0,$$

where $L(X, Y)$ is the set of all bounded linear operators from $X \to Y$.

Proposition 2.1.13. Let X, Y be two real Banach spaces and $\Omega \subset X$ be an open subset. If $F : \Omega \to Y$ is a continuous compact mapping and F is Fréchet differentiable at $x_0 \in \Omega$, then $F'(x_0)$ is compact.

Proof. Take any sequence $(h_n)_{n=1}^{\infty} \subset X$ with $\|h_n\| = 1$ for $n = 1, 2, \cdots$. Let $\epsilon > 0$. There exists $\delta_0 > 0$ such that $\frac{w(x_0, \delta h)}{\delta} < \frac{\epsilon}{2}$ for all $h \in X$ with $\|h\| = 1$ and $\delta \leq \delta_0$. Since F is compact, $(F(x_0 + \delta_0 h_n))$ has a convergent subsequence, say, $(F(x_0 + \delta_0 h_{n_k}))_{k=1}^{\infty}$. We also have

$$\begin{aligned} F'(x_0)h_{n_k} - F'(x_0)x_{n_l} = \delta_0^{-1}[F(x_0 + \delta_0 h_{n_k}) - F(x_0 + \delta_0 x_{n_l})] \\ + \delta_0^{-1}[w(x_0, \delta_0 h_{n_k}) - w(x_0, \delta_0 h_{n_l})]. \end{aligned}$$

Then we deduce, by letting $k, l \to \infty$, that

$$\|F'(x_0)h_{n_k} - F'(x_0)x_{n_l}\| \leq \epsilon.$$

Again, by letting $\epsilon \to 0$, we get that $(F'(x_0)h_{n_k})_{k=1}^{\infty}$ is convergent; thus $F'(x_0)$ is compact. This completes the proof.

Let E be a real Banach space and $C([0,T],E)$ be the space of continuous functions from $[0,T]$ to E with the norm $\|x(\cdot)\| = \max_{t\in[0,T]} \|x(t)\|$, then $C([0,T],E)$ is a Banach space. A subset $\mathcal{B} \subset C([0,T],E)$ is called *equicontinuous* if, for any $\epsilon > 0$, there exists $\delta(\epsilon) > 0$ such that $\|x(t_1) - x(t_2)\| < \epsilon$ for all $x(\cdot) \in \mathcal{B}$ and $t_1, t_2 \in [0,T]$ satisfying $|t_1 - t_2| < \delta$.

Theorem 2.1.14. (Ascoli Arzela's Theorem) A subset $M \subset C([0,T],E)$ is relatively compact if and only if
(1) M is equicontinuous and
(2) for each $t \in [0,T]$, $M(t) = \{x(t) : x(\cdot) \in M\}$ is relatively compact in E.

Proof. The proof follows easily from Theorem 3.1.16 in the next chapter.

Theorem 2.1.15. Let E be a real Banach space, $T : E \to E$ be a continuous compact mapping with $\lambda \in R$, $T_\lambda = T - \lambda I$ and $\sigma(T)$ be the spectrum of T. Then the following conclusions hold:

(1) If E is infinite dimensional, then $0 \in \sigma(T)$;

(2) If $0 \neq \lambda \in \sigma(T)$, then λ is an eigenvalue of T;

(3) $\sigma(T)$ is a countable subset;

(4) If $\lambda \notin \sigma(T)$, then T_λ is a homeomorphism onto X;

(5) $E = N(T_\lambda^k) \oplus R(T_\lambda^k)$ for all $\lambda \neq 0$, $k > 0$, where $dim(N(T_\lambda^k)) < +\infty$, and $R(N(T_\lambda^k))$ is closed in E.

Proof. (4) is obvious. For (1) suppose that $0 \notin \sigma(T)$. Then $T^{-1} : E \to E$ is continuous. Since E is infinite dimensional, there exists a sequence $(x_n)_{n=1}^{\infty} \subset$ such that $\|x_n\| = 1$ and $\|x_n - x_m\| \geq \frac{1}{2}$ for $n \neq m$. Put $y_n = Tx_n$, then (y_n) has a convergence subsequence, but $T^{-1}y_n = x_n$ will have a convergence subsequence, which is a contradiction. Thus $0 \in \sigma(T)$.

For (5) denote by $N_i = N(T_\lambda^i)$ and $R_i = R(T_\lambda^i)$ for $i = 1, 2, \cdots$. The finite dimensional property of N_i follows directly from the compactness of T. To prove the closedness of R_i, first, note there exists a closed subspace M of E such that $E = N_i \oplus M$. Define an operator $S : M \to E$ by

$$Sx = T_\lambda^i x \quad \text{for all } x \in M.$$

Observe that $R(S) = R(T_\lambda^i)$, so we only need to prove $R(S)$ is closed. It is obvious that S is one to one. Now, we prove that there exists $\gamma > 0$ such that

$$\|Sx\| \geq \gamma \|x\| \quad \text{for all } x \in M. \tag{2.1.1}$$

If this is not true, there exist $x_n \in M$ with $\|x_n\| = 1$ such that $\|Sx_n\| \leq n^{-1}$. Since $S = (T - \lambda I)^i = (-\lambda)^i I + \Sigma_{j=1}^i C_i^j (-\lambda)^{i-j} T^j$, one easily sees that $(x_n)_{n=1}^{\infty}$ has a convergence subsequence (x_{n_k}) with $x_{n_k} \to x_0$. Thus $Sx_0 = 0$, which implies that $x_0 \in N_i$ and so $x_0 = 0$, which contradicts $\|x_0\| = 1$. From (2.1.1), we deduce that $M = R(T_\lambda^i)$ is closed.

For (2) notice that $N_1 \subseteq N_2 \subseteq \cdots$ and $R_1 \supseteq R_2 \supseteq \cdots$. We cannot have $N_i \neq N_{i+1}$ for all i. Otherwise, by Lemma 2.1.9, there exists a sequence of $x_i \in N_{i+1} \setminus N_i$ such that $\|x_i\| = 1$ and $\|x_i - x_j\| \geq \frac{1}{2}$ for $i \neq j$ and thus we have

$$\|Tx_i - Tx_j\| \geq 2^{-1}\lambda \quad \text{for all} \quad j < i,$$

which is impossible since T is compact. Thus $N_j = N_i$ for some i and all $j > i$ and, consequently, $R_j = R_i$ for all $j > i$. If $\lambda \neq 0$ is not an eigenvalue of T, then T_λ is one to one. For any $y \in R_{i-1}$, we have $T_\lambda y \in R_i = R_{i+1}$ and thus there exists x such that $T_\lambda y = T_\lambda^{i+1} x$. Therefore, $y = T_\lambda^i x \in R_i$ and $R_{i-1} = R_i$, so on. Thus we get $R(T_\lambda) = E$. Therefore, $\lambda \notin \sigma(T)$.

To prove (3), we prove that, for any $r > 0$, $\{\lambda \in \sigma(T) : |\lambda| > r\}$ is a finite subset. Suppose the contrary, i.e., there exist $\lambda_n, |\lambda_n| > r$, $n = 1, 2, \cdots$, and $x_n \in E$ with $|x_n| = 1$ such that $Tx_n = \lambda_n x_n$ and $\{x_1, x_2, \cdots, x_n\}$ is linearly independent. Put $M_n = span\{x_1, x_2, \cdots, x_n\}$. Then $M_n \subset M_{n+1}$. By Lemma 2.1.9, there exist $y_{n+1} \in M_{n+1} \setminus M_n$ such that $\|y_{n+1}\| = 1$ and $d(y_{n+1}, M_n) \geq \frac{1}{2}$ for $n = 1, 2, \cdots$. Let $y_{n+1} = \Sigma_{i=1}^{n+1} \alpha_i^n x_i$. Then

$$\lambda_{n+1} y_{n+1} - A y_{n+1} = \Sigma_{i=1}^{n} \alpha_i^n (\lambda_{n+1} - \lambda_i) x_i = z_n \in M_n.$$

From which we deduce, for $m > n$, that

$$\begin{aligned} \|Ty_m - Ty_n\| &= \|(\lambda_m y_m - \lambda_n y_n) - (z_{m-1} - z_{n-1})\| \\ &\geq |\lambda_m| d(y_m, M_{m-1}) \geq \frac{r}{2}, \end{aligned}$$

which contradicts the compactness of T. Thus $\sigma(T)$ is countable. This completes the proof.

2.2 Leray Schauder Degree

In this section, we construct the Leray Schauder degree. First, we need the following result on the approximation of a compact mapping by finite dimensional mappings.

Lemma 2.2.1. Let E be a real Banach space, $\Omega \subset E$ be an open bounded subset and $T : \overline{\Omega} \to E$ be a continuous compact mapping. Then, for any $\epsilon > 0$, there exist a finite dimensional space F and a continuous mapping $T_\epsilon : \overline{\Omega} \to F$ such that

$$\|T_\epsilon x - Tx\| < \epsilon \quad \text{for all } x \in \overline{\Omega}.$$

Proof. Since $T\overline{\Omega}$ is relatively compact in E, for any $\epsilon > 0$, there exists a finite subset $\{x_1, x_2, \cdots, x_n\} \subset \overline{\Omega}$ such that

$$T(\overline{\Omega}) \subset \cup_{i=1}^{n} B(Tx_i, \epsilon).$$

Now, we define a mapping $T_\epsilon : \overline{\Omega} \to F = span\{Tx_1, Tx_2, \cdots, Tx_n\}$ as follows:

$$T_\epsilon x = \Sigma_{i=1}^{n} \frac{\phi_i(x)}{\Gamma(x)} Tx_i \quad \text{for all } x \in \overline{\Omega},$$

where $\phi_i(x) = \max\{0, \epsilon - \|Tx - Tx_i\|\}$ and $\Gamma(x) = \Sigma_{i=1}^{n} \phi_i(x)$. Then it is easy to check that T_ϵ satisfies the conditions of Lemma 2.2.1, so the conclusion follows. This completes the proof.

Lemma 2.2.2. Let E be a real Banach space, $B \subset E$ be a closed bounded subset and $T : B \to E$ be a continuous compact mapping. Suppose $Tx \neq x$ for all $x \in B$. Then there exists $\epsilon_0 > 0$ such that $x \neq tT_{\epsilon_1}x + (1-t)T_{\epsilon_2}x$ for all $t \in [0, 1]$ and $x \in B$, where $\epsilon_i \in (0, \epsilon_0)$ and $T_{\epsilon_i} : B \to F_{\epsilon_i}$ for $i = 1, 2$ as in Lemma 2.2.1.

Proof. Suppose the conclusion is not true. There exist $\epsilon_1^j \to 0$, $\epsilon_2^j \to 0$, $t_j \to t_0$, $x_j \in B$ such that $t_j T_{\epsilon_1^j} x_j + (1-t_j) T_{\epsilon_2^j} x_j = x_j$ for $j = 1, 2, \cdots$.

By compactness of T, $(Tx_j)_{j=1}^{\infty}$ has a subsequence, say (Tx_{j_k}), converging to $y \in E$. By Lemma 2.2.1, $T_{\epsilon_i^{j_k}} x_{j_k} \to y$ for $i = 1, 2$. Thus $x_{j_k} \to y \in B$. Therefore, $Ty = y$, which is a contradiction.

Definition 2.2.3. Let E be a real Banach space, $\Omega \subset E$ be an open bounded set and $T : \overline{\Omega} \to E$ be a continuous compact mapping. Now, suppose that $0 \notin (I - T)(\partial\Omega)$. Then, by Lemma 2.2.2, there exists $\epsilon_0 > 0$ such that

$$x \neq tT_{\epsilon_1}x + (1-t)T_{\epsilon_2}x \quad \text{for all } t \in [0, 1],\ x \in \partial\Omega,$$

where $\epsilon_i \in (0, \epsilon_0)$ and $T_{\epsilon_i} : \overline{\Omega} \to F_{\epsilon_i}$ for $i = 1, 2$ as in Lemma 2.2.1. Hence Brouwer's degree $deg(I - T_\epsilon, \Omega \cap F_\epsilon, 0)$ is well defined, and so we define

$$deg(I - T, \Omega, 0) = deg(I - T_\epsilon, \Omega \cap F_\epsilon, 0),$$

where $\epsilon \in (0, \epsilon_0)$.

By the homotopy property of Brouwer degree, we have

$$deg(I - T_{\epsilon_1}, \Omega \cap span\{F_{\epsilon_1} \cup F_{\epsilon_2}\}, 0) = deg(I - T_{\epsilon_2}, \Omega \cap span\{F_{\epsilon_1} \cup F_{\epsilon_2}\}, 0).$$

But $T_{\epsilon_i} : \overline{\Omega} \cap span\{F_{\epsilon_1} \cup F_{\epsilon_2}\} :\to F_i$ for $i = 1, 2$, so by Theorem 1.2.12 we have

$$deg(I - T_{\epsilon_1}, \Omega \cap span\{F_{\epsilon_1} \cup F_{\epsilon_2}\}, 0) = deg(I - T_{\epsilon_1}, \Omega \cap F_{\epsilon_1}, 0)$$

and

$$deg(I - T_{\epsilon_2}, \Omega \cap span\{F_{\epsilon_1} \cup F_{\epsilon_2}\}, 0) = deg(I - T_{\epsilon_2}, \Omega \cap F_{\epsilon_2}, 0).$$

Thus we have

$$deg(I - T_{\epsilon_1}, \Omega \cap F_{\epsilon_1}, 0) = deg(I - T_{\epsilon_2}, \Omega \cap F_{\epsilon_2}, 0)$$

and the degree defined in Definition 2.2.3 is well defined. For the general case, if $p \notin (I - T)(\partial\Omega)$, we define $deg(I - T, \Omega, p) = deg(I - T - p, \Omega, 0)$.

We recall some properties of the Leray Schauder degree as follows:

Theorem 2.2.4. The Leray Schauder degree has the following properties:

(1) (*Normality*) $deg(I, \Omega, 0) = 1$ if and only if $0 \in \Omega$;

(2) (*Solvability*) If $deg(I - T, \Omega, 0) \neq 0$, then $Tx = x$ has a solution in Ω,

(3) (*Homotopy*) Let $T_t : [0, 1] \times \overline{\Omega} \to E$ be continuous compact and $T_t x \neq x$ for all $(t, x) \in [0, 1] \times \partial\Omega$. Then $deg(I - T_t, \Omega, 0)$ doesn't depend on $t \in [0, 1]$;

(4) (*Additivity*) Let Ω_1, Ω_2 be two disjoint open subsets of Ω and $0 \notin (I - T)(\overline{\Omega} - \Omega_1 \cup \Omega_2)$. Then

$$deg(I - T, \Omega, 0) = deg(I - T, \Omega_1, 0) + deg(I - T, \Omega_2).$$

Proof. The proof follows from the corresponding properties of the Brouwer degree.

The following is the well-known Schauder fixed point theorem:

Theorem 2.2.5. Let $C \subset E$ be a nonempty bounded closed convex subset and $T : C \to C$ be a continuous compact mapping. Then T has a fixed point in C.

Proof. The proof is the same as the proof of Brouwer's fixed point theorem.

If we only require the continuous condition on the mapping T, then the conclusion of Theorem 2.2.5 fails as the following example shows:

Example 2.2.6. Let $T : l^2 \to l^2$ be a mapping defined by

$$T(x_1, x_2, \cdots) = (1 - \|x\|, x_1, x_2, \cdots)$$

for all $x = (x_1, x_2, \cdots) \in l^2$. Then $T : \overline{B(0, 1)} \to \overline{B(0, 1)}$ is continuous without a fixed point in $\overline{B(0, R)}$.

The Schauder fixed point theorem can be applied to yield the following result on hyper-invariant subspaces of a linear bounded operator:

Theorem 2.2.7. Let E be a Banach space, $T : E \to E$ be a nonzero linear continuous compact mapping and $\Gamma(T) = \{S \in L(E) : TS = ST\}$. Then there exists a nontrival hyper-invariant subspace F of T, i.e., $SF \subseteq F$ for all $S \in \Gamma(T)$.

Proof. Assume that the conclusion is not true. Then T does not have eignvalues. Take $x_0 \in E$ such that $\|x_0\| = 2$. Then, for any $x \in E$, we have $Tx \neq 0$ and $\{STx : S \in \Gamma(T)\}$ is a space which is invariant under $\Gamma(T)$ and so we have

$$\overline{\{Sx : S \in \Gamma(T)\}} = E.$$

Now, for any $y \in \overline{B(x_0, 1)}$, there exists $S_y \in \Gamma(T)$ such that $\|S_y Ty - x_0\| < 1$. By continuity of S, there exists $\delta(y) > 0$ such that $\|S_y Tx - x_0\| < 1$ for all $x \in \overline{B(x_0, 1)}$. Since $\{B(y, \delta(y)) : y \in \overline{B(x_0, 1)}\}$ is an open covering of $\overline{B(x_0, 1)}$, there exists a locally finite open refinement $\{V_i\}_{i\in I}$ of $\{B(y, \delta(y)) : y \in \overline{B(x_0, 1)}\}$.

Let $\{\phi_i\}_{i\in I}$ be a partition of unity subordinated to $\{V_i\}_{i\in I}$ and define a mapping $K : \overline{B(x_0, 1)} \to \overline{B(x_0, 1)}$ as follows:

$$Kx = \Sigma_{i\in I}\phi_i(x)S_iTx \quad \text{for all } x \in \overline{B(x_0, 1)}.$$

Since T is compact, K is a continuous compact mapping and hence T has a fixed point $y_0 \in \overline{B(x_0, 1)}$, i.e.,

$$\Sigma_{i\in I}\phi_i(y_0)S_iTy_0 = y_0.$$

Put $Z = \{y \in E : \Sigma_{i\in I}\phi_i(y_0)S_iTy = y\}$. Then Z is a finite dimensional subspace of E, and $T : Z \to Z$ and hence T has an eigenvalue, which is a contradiction. This completes the proof.

Theorem 2.2.8. Let E be a Banach space and $\Omega \subset E$ be an open bounded subset. If $T : \overline{\Omega} \to E$, $S : E \to E$ are continuous compact mappings and $p \notin (I - S)(I - T)(\partial\Omega)$, then

$$deg((I - S)(I - T), \Omega, p) = \Sigma_{i\in I} deg(I - T, \Omega, U_i) deg(I - S, U_i, p), \quad (2.2.1)$$

where $\{U_i\}_{i\in I}$ are connected component of $E\backslash(I-T)(\partial\Omega)$ and $deg(I-T, \Omega, U_i)$ is $deg(I - T, \Omega, z)$ for any $z \in U_i$.

Proof. We first prove that (2.2.1) only has finitely many nonzero terms. Take $r > 0$ such that $(I - T)(\overline{\Omega}) \subset B_r(0)$, then $M = \overline{B_r(0)} \cap (I - S)^{-1}(p)$ is compact, $M \subset R^n \setminus f(\partial\Omega) = \cup_{i\geq 1} U_i$ and there exists finitely many i, say $i = 1, 2, \cdots, t$, such that $\cup_{i=1}^{t+1} U_i \supseteq M$, where $U_{t+1} = U_\infty \cap B_{r+1}$.

We have $deg(I - T, \Omega, U_{t+1}) = 0$ and $deg(I - S, U_i, p) = 0$ for $i \geq t + 2$ since $U_j \subset B_r(0)$ and $g^{-1}(y) \cap U_j = \emptyset$ for $j \geq t + 2$. Therefore, the right side of (2.2.1) has only finitely many terms different from zero. Let

$$V_m = \{z \in B_{r+1}(0) \setminus (I - T)(\partial\Omega) : deg(I - T, \Omega, z) = m\}.$$

The same proof as in Theorem 1.2.13 yields

$$\Sigma_{i\in I} deg(I - T, \Omega, U_i) deg(I - S, U_i, p) = \Sigma_m m deg(I - S, V_m, p).$$

Now, we may choose ϵ sufficiently small. Let F be a finite dimensional subspace, $p \in F$ and $T_1 : \overline{\Omega} \to F$, $S_1 : \overline{B_{r+1}(0)} \to F$ be two continuous compact mappings such that

$$\|Tx - T_1x\| < \epsilon, x \in \overline{\Omega}, \|Sy - S_1y\| < \epsilon \quad \text{for all } y \in \overline{B_{r+1}(0)}.$$

Then, by the homotopy argument, we have

$$deg((I - S)(I - T), \Omega, p) = deg((I - S_1)(I - T_1), \Omega, p)$$

and

$$\begin{aligned} \Sigma_m m deg(I - S, V_m, p) &= \Sigma_m m deg(I - S_1, V_m, p) \\ &= \Sigma_m m deg(I - S_1, V_m \cap F, p). \end{aligned}$$

Therefore, by Theorem 1.2.13, we have

$$\Sigma_{i \in I} deg(I - T, \Omega, U_i) deg(I - S, U_i, p) = deg((I - S_1)(I - T_1), \Omega \cap F, p).$$

One easily gets

$$deg((I - S_1)(I - T_1), \Omega, p) = deg((I - S_1)(I - T_1), \Omega \cap F, p).$$

Thus the conclusion of Theorem 2.2.8 is true. This completes the proof.

Theorem 2.2.9. Let E be a Banach space, E_0 be a closed subspace of E and $\Omega \subset E$ be an open bounded subset. If $T : \overline{\Omega} \to E_0$ is a continuous compact mapping and $p \in E_0$, then $deg(I - T, \Omega, p) = deg(I - T, \Omega \cap E_0, p)$.

Proof. Since $T(\overline{\Omega}) \subset E_0$, we may choose a finite dimensional space $F \subset E_0$ with $p \in F$ and $T_1 : \overline{\Omega} \to F$ such that $\|Tx - T_1x\| < \epsilon$ in Definition 2.2.3 for small $\epsilon > 0$. Then we have

$$deg(I - T, \Omega, p) = deg(I - T_1, \Omega \cap F, p) = deg(I - T, \Omega \cap E_0, p).$$

Theorem 2.2.10. Let E be a Banach space and $0 \in \Omega \subset E$ with Ω be an open bounded subset. If $T : \overline{\Omega} \to E$ is a continuous compact mapping, then one of the following statements holds:

(1) T has a fixed point in $\overline{\Omega}$;

(2) There exist $\lambda > 1$ and $x \in \partial\Omega$ such that $Tx = \lambda x$.

Proof. If (2) holds, we are finished. Otherwise, put $H(t, x) = x - tTx$ for all $(t, x) \in [0, 1] \times \overline{\Omega}$. If $Tx = x$ for some $x \in \partial\Omega$, then (1) holds. Thus, we may assume that $Tx \neq x$ for all $x \in \partial\Omega$. Therefore, we have

$$x \notin tTx \quad \text{for all } (t, x) \in [0, 1] \times \partial\Omega.$$

By Theorem 2.2.4, we have

$$deg(I - T, \Omega, 0) = deg(I, \Omega, 0) = 1$$

and so T has a fixed point in Ω. This completes the proof.

Lemma 2.2.11. Let E be an infinite dimensional Banach space and $0 \notin \partial\Omega$ with Ω be an open bounded subset of E. Let $T : \overline{\Omega} \to E$ be a continuous compact mapping. Suppose that $Tx \neq \mu x$ for all $\mu \in [0,1]$, $x \in \partial\Omega$ and $0 \notin \overline{T\partial\Omega}$. Then $deg(I - T, \Omega, 0) = 0$.

Proof. First, we claim that there exists $\epsilon_0 > 0$ such that

$$\|Tx - T_\epsilon x\| < \epsilon, \quad \mu x \neq T_\epsilon x$$

for all $\mu \in [0,1]$, $x \in \partial\Omega$ and $\epsilon \in (0, \epsilon)$, where T_ϵ is the same as in Lemma 2.2.1.

If this is not true, there exist $\epsilon_j \to 0$, $x_j \in \partial\Omega$, $\mu_j \to \mu_0 \in [0,1]$ such that $\mu_j x_j = T_{\epsilon_j} x_j$ and so we have $Tx_j - \mu_j x_j \to 0$. Now, we have $0 \notin \overline{T\partial\Omega}$ and so $\mu_0 \neq 0$, (x_j) has a subsequence converging to $x_0 \in \partial\Omega$ and $Tx_0 = \mu_0 x_0$, which is a contradiction.

From the definition of the Leray Schauder degree, we know that

$$deg(I - T, \Omega, 0) = deg(I - T_\epsilon, \Omega \cap F, 0)$$

for sufficiently small ϵ and any $F \supset spanR(T_\epsilon)$. The homotopy invariance of Brouwer degree implies that

$$deg(I - T_\epsilon, \Omega \cap F, 0) = deg(-T_\epsilon, \Omega \cap F, 0).$$

Since E is an infinite dimensional Banach space, we may choose a finite dimensional subspace F of E such that $spanR(T_\epsilon)$ is a proper subspace of F, and $deg(-T_\epsilon, \Omega \cap F, 0) = deg(-T_\epsilon, \Omega \cap F, p)$ for any $p \in F$ with p sufficiently close to 0, so we must have $deg(-T_\epsilon, \Omega \cap F, 0) = 0$. Thus it follows that

$$deg(I - T, \Omega, 0) = 0.$$

This completes the proof.

Theorem 2.2.12. Let E be an infinite dimensional Banach space, $0 \in \Omega_0 \subset \Omega$ with Ω_0 and Ω be two open bounded subsets of E. Let $T : \overline{\Omega \setminus \Omega_0} \to E$ be a continuous compact mapping and suppose that the following conditions hold:

(1) $Tx \neq \lambda x$ for all $\lambda > 1$, $x \in \partial\Omega_0$,

(2) $Tx \neq \mu x$ for all $\mu \in [0,1)$, $x \in \partial\Omega$, and $0 \notin \overline{T\partial\Omega}$.

Then T has a fixed point in $\overline{\Omega \setminus \Omega_0}$.

Proof. We may assume that T is defined on $\overline{\Omega}$. Also, we assume that $Tx \neq x$ for all $x \in \partial\Omega_0 \cup \partial\Omega$. From (1) and Theorem 2.2.4, we have $deg(I - T, \Omega_0, 0) = 1$.

By (2) and Lemma 2.2.11, $deg(I - T, \Omega, 0) = 0$. Therefore, we have

$$deg(I - T, \Omega \setminus \overline{\Omega_0}, 0) = deg(I - T, \Omega, 0) - deg(I - T, \Omega_0, 0) = -1.$$

Consequently, T has a fixed point in $\Omega \setminus \overline{\Omega_0}$. This completes the proof.

From Theorem 2.2.12, we easily get the following result of Guo [141]:

Corollary 2.2.13. Let E be a infinite Banach space and $\Omega_0 \subset \Omega$ with $0 \in \Omega_0$, Ω be two open bounded subsets of E. Let $T : \overline{\Omega \setminus \Omega_0} \to E$ be a continuous compact mapping and, further, suppose that the following conditions hold:

(1) $\|Tx\| \leq \|x\|$ for all $x \in \partial\Omega_0$;

(2) $\|Tx\| \geq \|x\|$ for all $x \in \partial\Omega$.

Then T a fixed point in $\overline{\Omega \setminus \Omega_0}$.

Remark. The conclusion of Theorem 2.2.12 fails if we drop the infinite dimensional condition. The following example illustrates this:

Example 2.2.14. Let $\Omega = \{(x, y) : x^2+y^2 < 4\}$, $\Omega_1 = \{(x, y) : x^2+y^2 < 1\}$ and $T : \overline{\Omega \setminus \Omega_1} \to R^2$ be a rotation defined by

$$T(x, y) = \left(\sqrt{x^2 + y^2} \cos\left(\theta + \frac{\pi}{4}\right), \sqrt{x^2 + y^2} \sin\left(\theta + \frac{\pi}{4}\right)\right)$$

for all $(x, y) \in \overline{\Omega \setminus \Omega_1}$, where $x + yi = \sqrt{x^2 + y^2} e^{i\theta}$. Then we have $T(x, y) \neq \mu(x, y)$ for all $\mu \in [0, 1]$, $x^2 + y^2 = 1$ and $T(x, y) \neq \lambda(x, y)$ for all $\lambda \geq 1$, $x^2 + y^2 = 4$. However, the mapping T does not have a fixed point in $\overline{\Omega \setminus \Omega_1}$.

We have the following result in any dimensional Banach spaces:

Theorem 2.2.15. Let E be a Banach space and $0 \in \Omega_0 \subset \Omega$ with Ω_0, Ω two open bounded subsets of E. Let $T : \overline{\Omega \setminus \Omega_0} \to E$ be a continuous compact mapping and, further, suppose that the following conditions hold:

(1) $Tx \neq \lambda x$ for all $\lambda > 1$ and $x \in \partial\Omega_0$;

(2) $Tx \neq \mu x$ for all $\mu \in [0, 1)$, $x \in \partial\Omega$ and $0 \notin Conv\overline{T\partial\Omega}$.

Then T has a fixed point in $\overline{\Omega \setminus \Omega_0}$.

Proof. As in the proof of Theorem 2.2.12, we may assume that $Tx \neq x$ for $x \in \partial\Omega_0 \cup \partial\Omega$. We only need to show that

$$deg(I - T, \Omega, 0) = 0.$$

Assume that this is not true. Then there exists a compact mapping $T_1 : \overline{\Omega} \to Conv\overline{T\partial\Omega}$ such that $T_1 x = Tx$ for $x \in \partial\Omega$. For $k > 1$, it is easy to see that

$$x \neq tTx + (1-t)kT_1 x \quad \text{for all } (t, x) \in [0, 1] \times \partial\Omega.$$

Thus we have

$$deg(I - kT_1, \Omega, 0) = deg(I - T, \Omega, 0) \neq 0$$

and so $kT_1 x = x$ has a solution in Ω for $k > 1$, which contradicts the fact that Ω is unbounded. Hence we have $deg(I - T, \Omega, 0) = 1$. This completes the proof.

Theorem 2.2.16. Let E be a Banach space, Ω with $0 \in \Omega$ be an open bounded subset of E and $L : \overline{\Omega} \to E$ be a linear continuous compact mapping. If $\lambda \neq 0$ and λ^{-1} is not an eigenvalue of L, then

$$deg(I - \lambda L, \Omega, 0) = (-1)^{m(\lambda)},$$

where $m(\lambda)$ is the sum of the algebraic multiplicities of the eigenvalues μ satisfying $\mu\lambda > 1$, and, if L has no such eigenvalues μ, then $m(\lambda) = 0$.

Proof. Put $S = I - \lambda L$, then S is a homeomorphism onto E. There are at most finitely many eigenvalues of L such that $\mu\lambda > 1$, say, μ_i, $i = 1, 2, \cdots, k$. Set $F = \oplus_{i=1}^k N(\mu_i)$ and $W = \cap_{i=1}^k R(\mu_i)$, where $N(\mu_i) = \{x : Lx - \mu_i x = 0\}$ and $R(\mu_i) = (L - \mu_i)(E)$ for $i = 1, 2, \cdots, k$. We know from the spectral theory of linear compact mappings that $N(\mu_i)$ are finitely dimensional spaces, $i = 1, 2, \cdots, k$. It is easy to see that $E = F \oplus W$. There are projections $P : E \to F$, and $Q : E \to W$. Set $L_1 = SP + Q$ and $L_2 = P + SQ$. Then we have

$$I - L_1 = -\lambda LP_1, \quad I - L_2 = -\lambda LQ, \quad S = L_1 L_2,$$
$$(I - L_1)(F) \subset F, \quad (I - L_2)(W) \subset W.$$

Moreover, L_i is one to one for $i = 1, 2$. Thus, by product formula, we have

$$deg(S, \Omega, 0) = deg(L_1 L_2, \Omega, 0) = deg(L_1, \Omega, 0) deg(L_2, \Omega, 0).$$

By Theorem 2.2.9, we have

$$deg(L_1, \Omega, 0) = deg(L_1, \Omega \cap F, 0)$$

and

$$deg((L_2, \Omega, 0) = deg(L_2, \Omega \cap W, 0).$$

But $I - t\lambda L$ has no solution on $\partial\Omega \cap W$ and so we have

$$deg(L_2, \Omega \cap W, 0) = deg(I, \Omega \cap W, 0) = 1.$$

On the other hand, the eigenvalues of L_1 are the eigenvalues of $I - \lambda L$ on F, i.e., $1 - \lambda\mu_i$ for $i = 1, 2, \cdots, k$. Thus we have

$$deg(L_1, \Omega \cap F, 0) = sgndet(L_1|_F) = sgn\Pi_{i=1}^k (1 - \lambda\mu_i)^{dim(N(\mu_i)}$$
$$= (-)^{\Sigma_{i=1}^k dim(N(\mu_i))} = (-1)^{m(\lambda)}.$$

2.3 Leray Schauder Degree for Multi-Valued Mappings

In this section, we describe the Leray Schauder degree for upper semicontinuous compact mapping with closed convex values. First, we introduce several multi-valued maps, which play very important roles in the study of nonsmooth analysis, differential inclusions and nonlinear partial differential equations. We begin with the following definitions:

Definition 2.3.1. Let X, Y be two topological spaces and $T : X \to 2^Y$ be a multi-valued mapping, i.e., Tx is a subset of Y for all $x \in X$.

(1) T is said to be lower semicontinuous at x_0 if, for any open set V of Y with $Tx_0 \cap V \neq \emptyset$, the set $U = \{x : Tx \cap V \neq \emptyset\}$ is open in X, and if T is lower semicontinuous at every point of X, T is said to be lower semicontinuous on X,

(2) T is said to be upper semicontinuous at x_0 if, for any open neighborhood $V(Tx_0)$ of Tx_0, there exists an open neighborhood of $U(x_0)$ such that $TU \subset V$, and T is said to be upper semicontinuous on X if T is upper semicontinuous at every point of X,

(3) T is said to be continuous at x_0 if T is both upper semicontinuous and lower semicontinuous at x_0, and if T is continuous at all points of X, then T is said to be continuous on X.

Definition 2.3.2. Let X be a topological space and Y be a normed space. A mapping $T : X \to 2^Y$ is called Hausdorff continuous at x_0 if

$$\lim_{x \to x_0} H(Tx, Tx_0) = 0$$

and, if T is Hausdorff continuous at all points of X, then T is said to be Hausdorff continuous on X.

Next, we give some examples of multi-valued mappings, which have appeared in several fields of mathematics.

Example 2.3.3. Let $f : R^n \to R^n$ be a bounded function (not necessarily continuous). We define a multi-valued mapping $F : R^n \to 2^{R^n}$ as follows:

$$Fx = \{y : \text{ there exist } x_j \in R^n \text{ such that } x_j \to x, fx_j \to y\}.$$

Then F is upper semicontinuous. Such a mapping has been used to study differential equations with discontinuous right-hand sides (see [12], [114]).

Example 2.3.4. Let X be a Banach space and $f : X \to R$ be a locally Lipschitz function. The Clarke derivative of f at x in the direction v is defined by

$$f^o(x; v) = \limsup_{t \downarrow 0, y \to x} \frac{f(y + tv) - f(y)}{t}$$

and the Clarke subdifferential of f at x is defined by

$$\partial f(x) = \{\phi \in X^* : \phi(v) \leq f^o(x; v)\}.$$

This subdifferential is a multi-valued mapping and it is a powerful tool in non-smooth analysis (see [72]).

Example 2.3.5. Let $u : R^n \to R$ be a continuous function. We define

$$D^+u(x) = \Big\{p \in R^N : \limsup_{y \to x, y \in \Omega} \frac{u(y) - u(x) - p(y-x)}{|y-x|} \leq 0\Big\},$$

$$D^-u(x) = \Big\{p \in R^N : \liminf_{y \to x, y \in \Omega} \frac{u(y) - u(x) - p(y-x)}{|y-x|} \geq 0\Big\}.$$

D^+u and D^-u are used to define the viscosity solutions for fully nonlinear partial differential equations (see [75]).

Example 2.3.6. Let X be a infinite Banach space, $r > 0$ be a constant and $F : X \to 2^X$ be defined as follows:

$$Fx = B(x, r) \quad \text{for all } x \in X.$$

Then F is lower semicontinuous on X, but not upper semicontinuous.

Indeed, for any $x_0 \in X$ and $y \in B(x_0, r)$, we have $(y - x_0) + x \in F(x) = B(x, r)$ and so $(y - x_0) + x \to y$ as $x \to x_0$. Therefore, F is lower semicontinuous at x_0. To see that F is not upper semicontinuous, we know, since X is infinite dimensional, that there exists a sequence $\{x_n\} \subset X$ such that

$$\|x_n\| = 1, \quad \|x_n - x_m\| \geq \epsilon,$$

where $\epsilon \in (0, 1)$ is a constant. Then we have $(r + \frac{1}{n})x_n + x_0 \in F(x_0 + \frac{1}{n}x_n)$, but $V = X \setminus \{(r + \frac{1}{n})x_n + x_0 : n = 1, 2 \cdots,\}$ is open and $F(x_0, r) \subset V$. Therefore, $F(x_0 + \frac{1}{n}x_n) \not\subset V$, i.e., F is not upper semicontinuous. Moreover, it is easy to check that F is Hausdorff continuous on X.

Lemma 2.3.7. Let X be a metric space, Y be a normed space and $T : X \to 2^Y$ be an upper semicontinuous mapping with closed convex values. Then, for any $\epsilon > 0$, there exists a continuous mapping $f_\epsilon : X \to conv(TX)$ such that, for any $x \in X$, there exist $y \in X$ and $z \in Ty$ such that

$$d(x, y) < \epsilon, \quad \|f_\epsilon x - z\| < \epsilon.$$

Proof. For any $x \in X$ and $\epsilon > 0$, there exists $\delta_x > 0$ such that

$$TB(x, \delta_x) \subset B(Tx, \epsilon).$$

We may require $\delta_x < \epsilon$. Let $\{U_i\}_{i \in I}$ be a locally finite open refinement of $\{B(x, \frac{\delta_x}{2}) : x \in X\}$ and $\{\phi_i\}_{i \in I}$ be a partition of the unity subordinated $\{U_i\}_{i \in I}$.

Now, we define a mapping $f_\epsilon : X \to Y$ as follows:

$$f_\epsilon x = \Sigma_{i\in I}\phi_i(x)y_i \quad \text{for all } x \in X,$$

where $U_i \subset B(x_i, \frac{\delta_x}{2})$ and $y_i \in Tx_i$. It is obvious that $f_\epsilon : X \to conv(TX)$ is continuous. For any given $x \in X$, let $I_0 = \{i \in I : \phi_i(x) \neq 0\}$. Then there exists $i_0 \in I_0$ such that

$$\delta_{x_{i_0}} = \max_{i\in I_0}\{\delta_{x_i}\}.$$

We put $y = x_{i_0}$. For $i \in I_0$, we have $x \in U_i \subset B(x_i, \frac{\delta_{x_i}}{2})$ and hence $x_i \in B(x_{i_0}, \delta_{x_{i_0}})$. Therefore, we have

$$f_\epsilon x = \Sigma_{i\in I_0}\phi_i(x)y_i \in B(Ty, \epsilon).$$

Take $z \in Ty$ such that $\|f_\epsilon x - z\| < \epsilon$. This completes the proof.

Theorem 2.3.8. Let X be a metric space, Y be a Banach space and $T : X \to 2^Y$ be a lower semicontinuous mapping with closed convex values. Then there exists a single-valued continuous mapping $f : X \to Y$ such that $f(x) \in Tx$ for all $x \in X$.

Proof. First, we prove that, for any $\epsilon > 0$, there exists a continuous mapping $f : X \to Y$ such that $f(x) \in Tx + B_\epsilon(0)$ for all $x \in X$. To see this, if we put $U(y) = \{x \in X : x \in T^{-1}(y - B_\epsilon)\}$, then $U(y)$ is open in X by the lower semicontinuity of T. Now, since we have $X = \cup_{y\in Y}U(y)$, there is a locally finite open refinement $\{W_i\}_{i\in I}$ of $\{U(y)\}_{y\in Y}$ and a partition of unity $\{\alpha_i\}_{i\in I}$ subordinated to $\{W_i\}_{i\in I}$. Pick y_i such that $W_i \subset U(y_i)$ and define a mapping $f : X \to Y$ as follows:

$$f(x) = \Sigma_{i\in I}\alpha_i(x)y_i \quad \text{for all } x \in X.$$

Then f is continuous. Obviously, if $\alpha_i(x) \neq 0$, then $x \in W_i \subset U(y_i)$ and so we have $y_i \in Fx + B_\epsilon(0)$. Thus it follows that

$$f(x) \in F(x) + B_\epsilon(0)$$

since the right hand side is convex. Now, for $\epsilon_n = 2^{-n}$ for $n = 1, 2, \cdots$, by the above conclusion, we have a sequence $\{f_n\}$ of continuous mappings such that

$$f_n \in f_{n-1}(x) + 2B_{\epsilon_n}(0), \quad f_n(x) \in Tx + B_{\epsilon_n}(0)$$

for all $x \in X$. Indeed, assume that we have defined mappings $f_1, f_2, \cdots, f_n$ for some $n > 1$, respectively, and put $Gx = Tx \cap (f_n + B_\epsilon(0))$ for all $x \in X$. Then G is lower semicontinuous with convex values.

By the first step, there exists $f_{n+1} : X \to Y$ such that $f_{n+1}(x) \in G(x) + B_{\epsilon_{n+1}}$ and so we have

$$f_{n+1}(x) \in T(x) + B_{\epsilon_{n+1}}, \quad f_{n+1}(x) \in f_n(x) + 2B_{\epsilon_n}.$$

Evidently, since $\{f_n\}$ is a Cauchy sequence, let $f(x) = \lim_{n\to\infty} f_n(x)$. Then f is continuous since the convergence is uniform and $f(x) \in Tx$ for all $x \in X$. This completes the proof.

In the above result, we considered approximate selections and continuous selections of multi-valued mappings. In some cases, we need to find a measurable selection for a given multi-valued mapping. Let $(\Omega, \mathcal{A})$ be a measurable space, (X, d) be a separable metric space and $f : \Omega \to 2^X$ be a multi-valued mapping. Then f is called a measurable function if $f^{-1}(B) \in \mathcal{A}$ for all open subset $B \subset X$.

Theorem 2.3.9. Let $(\Omega, \mathcal{A})$ be a measurable space, (X, d) be a separable complete metric space and $F : \Omega \to 2^X$ be a measurable multi-valued mapping with closed values. Then there exists a single valued measurable mapping $f : \Omega \to X$ such that $f(x) \in Fx$ for all $x \in \Omega$.

Proof. Since X is separable, there exists a countable subset $\{x_1, x_2, \cdots\}$ of X such that

$$\overline{\{x_1, x_2, \cdots\}} = X.$$

We shall define a sequence $\{f_n\}$ of measurable functions satisfying the following:

(1) $d(f_n(x), f_{n+1}(x)) < 2^{-n+1}$ for all $x \in \Omega$;

(2) $d(f_n(x), F(x)) < 2^{-n}$ for all $x \in \Omega$

and define a mapping $f_1 : \Omega \to X$ by $f_1(z) = x_k$ if k is the smallest integer such that

$$Fz \cap B(x_k, 1) \neq \emptyset.$$

Since $f^{-1}(x_k) = F^{-1}(B(x_k, 1) \setminus \cup_{m<k} F^{-1}(B(x_m, 1))$, it follows that f_1 is measurable.

Now, assume that we have defined f_1, f_2, $\cdots$, f_k satisfying (1) and (2) for some $k > 1$. For each $z \in \Omega$, we have $z \in f_k^{-1}(x_i)$ for some i. Now, if $z \in f_k^{-1}(x_i)$, then we define

$$f_{k+1}(z) = x_p,$$

where p is the smallest integer such that

$$Fz \cap B(x_i, 2^{-k}) \cap B(x_p, 2^{-k-1}) \neq \emptyset.$$

Obviously, f_{k+1} is well defined and it is also measurable and, moreover, it satisfies (1) and (2).

Finally, if we put $f(x) = \lim_{n\to\infty} f_n(x)$, then f is measurable and $f(x) \in Fx$ for all $x \in \Omega$. This completes the proof.

Example 2.3.10. Let $F : R^n \to 2^R \setminus \emptyset$ be a upper semicontinuous function with bounded closed values. Assume that $|y| \leq M|x| + f(x)$ for all $y \in Fx$

and $x \in R^n$, where $M > 0$ is a constant and $f(\cdot) \in L^p(R^n)$ for $p \in [1, \infty)$. We define a mapping $\mathcal{F} : C([0,1]; R^n) \to L^p(R^n)$ by

$$\mathcal{F}u(t) = \{g(\cdot) \in L^p([0,1]) : g(t) \in Fu(t),\ a.e. t \in [0,1]\}$$

for all $u(\cdot) \in C([0,1]; R^n)$. Since $Fu(t) : [0,1] \to 2^R$ is upper semicontinuous with closed values, it is measurable. By Theorem 2.3.9, we know that there exists a measurable selection $g(t) \in Fu(t)$ and, by assumption, we have

$$|g(t)| \leq M|u(t)| + f(u(t)) \quad \text{for all } t \in [0,1].$$

Therefore, it follows that $g(\cdot) \in L^p([0,1])$ and hence $\mathcal{F}$ is well defined.

Next, we show that the Leray Schauder degree can be generalized to multi-valued upper semicontinuous compact mappings with closed convex values:

Proposition 2.3.11. Let E be a real Banach space, $\Omega \subset E$ be an open bounded set and $T : \overline{\Omega} \to 2^E$ be an upper semicontinuous mapping with closed convex values. If $T\overline{\Omega}$ is relatively compact and $x \notin Tx$ for all $x \in \partial\Omega$, then there exists $\epsilon_0 > 0$ such that $x \neq f_\epsilon x$ for all $x \in \partial\Omega$ and $\epsilon \in (0, \epsilon_0)$, where f_ϵ is defined as in Lemma 2.3.7.

Proof. Suppose that the conclusion is not true. Then there exist $\epsilon_j \to 0$ and $x_j \in \partial\Omega$ such that $x_j = f_{\epsilon_j} x_j$. By Lemma 2.3.7, there exist $y_j \in \overline{\Omega}$ and $z_j \in Ty_j$ such that

$$\|x_j - y_j\| \leq \epsilon_j, \quad \|f_{\epsilon_j} x_j - z_j\| < \epsilon_j.$$

Since T is compact, we may assume that $z_j \to z_0$ and hence we have

$$f_{\epsilon_j} x_j \to z_0, \quad x_j \to z_0 \in \partial\Omega$$

and thus $y_j \to z_0$. Using the upper semicontinuity of T, we get $z_0 \in Tz_0$, which is a contradiction. This completes the proof.

Definition 2.3.12. Let E be a real Banach space, $\Omega \subset E$ be an open bounded set and $T : \overline{\Omega} \to 2^E$ be an upper semicontinuous mapping with closed convex values. Suppose that $T\overline{\Omega}$ is relatively compact and $x \notin Tx$ for all $x \in \partial\Omega$. Then we define

$$deg(I - T, \Omega, 0) = \lim_{\epsilon \to 0} deg(I - f_\epsilon, \Omega, 0),$$

where f_ϵ is defined as in Lemma 2.3.7.

We show that Definition 2.3.12 is reasonable. By Proposition 2.3.11, there exists $\epsilon_0 > 0$ such that $x \neq f_\epsilon x$ for all $x \in \partial\Omega$ and $\epsilon \in (0, \epsilon_0)$. It is obvious that $f_\epsilon(\overline{\Omega})$ is compact. Thus $deg(I - f_\epsilon, \Omega, 0)$ is well defined for all $\epsilon \in (0, \epsilon_0)$. We claim that there exists $\epsilon_1 < \epsilon_0$ such that $tf_\epsilon x + (1-t)f_\delta x \neq x$ for all

$(t,x) \in [0,1] \times \partial\Omega$ and $\epsilon, \delta \in (0, \epsilon_1)$. If not, there exist $t_j \to t_0$, $\epsilon_j \to 0$, $\delta_j \to 0$, and $x_j \in \partial\Omega$ such that

$$t_j f_{\epsilon_j} x_j + (1-t_j) f_{\delta_j} x_j = x_j.$$

By Lemma 2.3.7, there exist y_j^1, y_j^2 and $z_j^1 \in Ty_j^1, z_j^2 \in Ty_j^2$ satisfying

$$\|y_j^1 - x_j\| < \epsilon_j, \ \|y_j^2 - x_j\| < \delta_j, \ \|z_j^1 - f_{\epsilon_j} x_j\| < \epsilon_j, \ \|z_j^2 - f_{\delta_j} x_j\| < \delta_j.$$

By compactness of T, we may assume that $x_j \to x_0 \in \partial\Omega$. Therefore, $y_j^1 \to x_0$ and $y_j^2 \to x_0$. Consequently, (z_j^1) has a subsequence $(z_{j_k}^1)$ with $z_{j_k}^1 \to z_1 \in Tx_0$, (z_j^2) has a subsequence $(z_{j_k}^2)$ with $z_{j_k}^2 \to z_2 \in Tx_0$ and $tz_1 + (1-t)z_2 = x_0 \in Tx_0$, which is a contradiction.

The following property follows from Theorem 2.2.4, and we leave the proofs to the reader:

Theorem 2.3.13. The degree defined by Definition 2.3.12 has the following properties:

(1) (*Normality*) $deg(I, \Omega, 0) = 1$ if and only if $0 \in \Omega$;

(2) (*Solvability*) If $deg(I - T, \Omega, 0) \neq 0$, then $x \in Tx$ has a solution in Ω;

(3) (*Homotopy*) Let $T_t : [0,1] \times \overline{\Omega} \to E$ be a upper semicontinuous compact mapping with closed convex values and $x \notin T_t x$ for all $(t,x) \in [0,1] \times \partial\Omega$. Then $deg(I - T_t, \Omega, 0)$ does not depend on $t \in [0,1]$;

(4) (*Additivity*) If Ω_1, Ω_2 are two disjoint open subsets of Ω and $0 \notin (I - T)(\overline{\Omega} - \Omega_1 \cup \Omega_2)$, then

$$deg(I - T, \Omega, 0) = deg(I - T, \Omega_1, 0) + deg(I - T, \Omega_2).$$

2.4 Applications to Bifurcations

In this section, we give some applications to bifurcation.

Definition 2.4.1. Let X, Y be two Banach spaces, $\alpha_0 > 0$, $\delta > 0$, $I = (-\alpha_0 - \delta, \alpha_0 + \delta)$, $x_0 \in \Omega \subset X$ be an open subset and $F : I \times \Omega \to Y$ be such that $F(\alpha, x_0) = 0$ on I. If there exists $\alpha_n \to \alpha_0$, $x_n \in \Omega \setminus \{x_0\}, x_n \to x_0$ such that $F(\alpha_n, x_n) = 0$, then we call (α_0, x_0) a bifurcation point for $F(\alpha, x) = 0$.

Theorem 2.4.2. Let X be an infinite dimensional Banach space, $\Omega \subset X$ be an open subset with $0 \in \Omega$ and $F : \Omega \to X$ be a continuous compact operator with $F0 = 0$. Suppose $\liminf_{x \to 0} \frac{\|Fx\|}{\|x\|} = +\infty$. Then $(0,0)$ is a bifurcation point for $x - \alpha Fx = 0$.

Proof. Take $\alpha_n \to 0$. Consider the mapping $H(t,x) = x - \alpha_n tFx$ for $(t,x) \in [0,1] \times \Omega$. Since $\liminf_{x\to 0} \frac{\|Fx\|}{\|x\|} = +\infty$, there exist $r_n < \frac{1}{n}$ such that

$$\|Fx\| > \alpha_n^{-1}\|x\|, x \in \partial B(0,r_n).$$

By Lemma 2.2.11, we have $deg(I - \alpha_n F, B(0,r_n), 0) = 0$. However, it follows that

$$deg(I, B(0,r_n), 0) = 1,$$

so there must exist $t_n \in (0,1)$, $x_n \in \partial B(0,r_n)$ such that $H(t_n,x_n) = 0$, i.e., $x_n - \alpha_n t_n Fx_n = 0$. Thus $(0,0)$ is a bifurcation point. This completes the proof.

Lemma 2.4.3. Let (M,d) be a compact metric space, $A_1 \subset M$ be a component and $A_2 \subset M$ closed such that $A_1 \cap A_2 = \emptyset$. Then there exist compact sets M_1, M_2 such that $A_i \subset M_i$, $i = 1,2$, $M_1 \cap M_2 = \emptyset$ and $M = M_1 \cup M_2$.

Proof. For $\epsilon > 0$, $a, b \in M$, if there exist finitely many $x_i \in M$, $i = 1, 2, \cdots, n$, such that $x_1 = a, x_n = b$ and $d(x_{i+1}, x_i) < \epsilon$ for $i = 1, 2, \cdots, n-1$, then we call a, b ϵ chainable. Put

$$A_\epsilon = \{x \in M: \text{ there exists } a \in A_1 \text{ such that } x \text{ and } aare\epsilon \text{ chainable}\}.$$

Clearly, $A_1 \subset A_\epsilon$ and A_ϵ is both open and closed in M.

Now, we prove that there exists $\epsilon_0 > 0$ such that $A_{\epsilon_0} \cap A_2 = \emptyset$. If this is not true, for $\epsilon_j \to 0$, there exist $y_j \in A_{\epsilon_j} \cap A_2$. So there exist $a_j \in A_1$ such that y_j, a_j are ϵ_j-chainable. However, A_1, A_2 are compact, so we may assume that $a_j \to x_0$, $b_j \to y_0$. Consequently, we have ϵ_j chains joining x_0, y_0 for every $j \geq 1$. We set

$$C = \{x \in m: \text{ there exist } j_k \to \infty, x_{j_k} \in M_{j_k} \text{ such that } x_{j_k} \to x\}.$$

Obviously, C is compact and $x_0, y_0 \in C$. Notice that any two points in C are ϵ_j-chainable, so C must be connected. However, $C \cap A_1 \neq \emptyset$, so we have $C \subset A_1$, and thus $y_0 \in A_1 \cap A_2$, which is a contradiction. Therefore, there exist $\epsilon_0 > 0$ such that $A_{\epsilon_0} \cap A_2 = \emptyset$. Set $M_1 = A_{\epsilon_0}$ and $M_2 = M \setminus M_1$, then we get the desired result.

Theorem 2.4.4. Let X be a real Banach space, $\Omega \subset R \times X$ be a open neighborhood of $(\alpha_0, 0)$ and $T : \overline{\Omega} \to X$ be a continuous compact mapping with $T(\alpha, x) = o(\|x\|)$ as $x \to 0$ uniformly in α. Let $S : X \to X$ be a linear continuous compact mapping, α_0^{-1} be an eigenvalue of odd algebraic multiplicity and

$$M = \{(\alpha, x) \in \Omega : x - \alpha Sx + T(\alpha, x) = 0, \ x \neq 0\}.$$

Then the component C of $\overline{M}$ containing $(\alpha_0, 0)$ has at least one of the following properties:

(1) $C \cap \partial\Omega \neq \emptyset$.

(2) C contains an odd number of trivial zeros $(\alpha_i, 0) \neq (\alpha_0, 0)$, where α_i^{-1} is an eigenvalue of S of odd algebraic multiplicity.

Proof. First, if $C \cap \partial\Omega \neq \emptyset$, then C is compact and contains another $(\alpha, 0)$ with $\alpha \neq \alpha_0$. The compactness of C follows from the compactness of S and T. Suppose $C \cap R \times \{0\} = \{(\alpha_0, 0)\}$. For any $\delta > 0$, set

$$N_\delta = \{(\alpha, x) \in \Omega : d((\alpha, x), C) < \delta\}.$$

If $\overline{N_\delta} \cap \overline{M} = C$, we put $\Omega_0 = N_\delta$. Otherwise, by Lemma 2.4.3, there exist compact sets C_1, C_2 such that $C \subset C_1$, $\overline{M} \cap \partial N_\delta \subset C_2$, $C_1 \cap C_2 = \emptyset$ and $C_1 \cup C_2 = \overline{N_\delta} \cap \overline{M}$. Set $\gamma = d(C_1, C_2)$, then $\gamma > 0$. We put $\Omega_0 = N_\delta \cap \{x : d(x, C_1) < \frac{\gamma}{2}\}$. It is obvious that

$$C \subset \Omega_0 \subset \overline{\Omega_0} \subset \Omega, \quad \overline{M} \cap \partial\Omega_0 = \emptyset.$$

Now, we may take $\delta > 0$ small enough such that no other eigenvalue α^{-1} of S satisfies $|\alpha - \alpha_0| \leq 2\delta$ and the intersection of $\overline{M}$ and the real line is given by $I = [\alpha_0 - \delta, \alpha_0 + \delta]$. Since $\overline{M} \cap \partial\Omega_0 = \emptyset$, $deg(I - \alpha S - T(\alpha, \cdot), \Omega(\alpha), 0)$ is constant on I, where $\Omega(\alpha) = \{x : (\alpha, x) \in \Omega_0\}$.

Now, we choose $\alpha_0 - \delta < \alpha_1 < \alpha_0 < \alpha_0 + \delta$. For r sufficiently small, we have

$$\begin{aligned} &deg(I - \alpha_i S - T(\alpha_i, \cdot), \Omega(\alpha_i), 0) \\ &= deg(I - \alpha_i S - T(\alpha_i, \cdot), \Omega(\alpha_i) \setminus \overline{B(0, r)}, 0) \\ &\quad + deg(I - \alpha_i S - T(\alpha_i, \cdot), B(0, r), 0) \end{aligned}$$

for $i = 1, 2$. However, $deg(I - \alpha_1 S, T(\alpha_1, \cdot), B(0, r), 0)$ is different to $deg(I - \alpha_2 S, T(\alpha_2, \cdot), B(0, r)\Omega, 0)$ by a factor -1 and $deg(I - \alpha_i S - T(\alpha_i, \cdot), \Omega(\alpha_i) \setminus B(0, r), 0) = 0$, which is a contradiction. Clearly, a bounded open neighborhood Ω_0 of C satisfying $\overline{M} \cap \partial\Omega_0 = \emptyset$ contains a finite number of points $(\alpha_i, 0)$ with $\alpha_i^{-1} \in \sigma(S)$, say, $\alpha_1 < \cdots < \alpha_{j-1} < \alpha_0 < \alpha_{j+1} < \cdots < \alpha_N$. We may take $\delta > 0$ sufficiently small such that

$$\overline{\Omega_0} \cap R \times \{0\} = (\cup_{i=1}^N [\alpha_i - \delta, \alpha_i + \delta]) \times \{0\}.$$

Choose α_{j1}, α_{j2} such that

$$\alpha_j - \delta < \alpha_{j1} < \alpha_j < \alpha_{j2} < \alpha_j + \delta.$$

We have $deg(I - \alpha S - T(\alpha, \cdot), \Omega_0(\alpha), 0) = m$ on $[\alpha_1 - \delta, \alpha_N + \delta]$ for some $m \in Z$ and

$$\begin{aligned} m &= deg(I - \alpha_{ji} S - T(\alpha_{ji}, \cdot), \Omega_0(\alpha_{ji}), 0) \\ &= deg(I - \alpha_{ji} S - T(\alpha_{ji}, \cdot), \Omega_0(\alpha_{ji}) \setminus \overline{B(0, r)}, 0) \\ &\quad + deg(I - \alpha_{ji} S - T(\alpha_{ji}, \cdot), B(0, r), 0) \end{aligned}$$

for $i = 1, 2$, where $r > 0$ is sufficiently small.

Moreover, it follows that

$$\begin{aligned} &deg(I - \alpha_{11}S - T(\alpha_{11}, \cdot), \Omega_0(\alpha_{11}) \setminus \overline{B(0,r)}, 0) \\ &= deg(I - \alpha_{N2}S - T(\alpha_{N2}, \cdot), \Omega_0(\alpha_{N2}) \setminus \overline{B(0,r)}, 0) \\ &= 0 \end{aligned}$$

and

$$\begin{aligned} &deg(I - \alpha_{j2}S - T(\alpha_{j2}, \cdot), \Omega_0(\alpha_{j2}) \setminus \overline{B(0,r)}, 0) \\ &= deg(I - \alpha_{(j+1)1}S - T(\alpha_{(j+1)1}, \cdot), \Omega_0(\alpha_{(j+1)1}) \setminus \overline{B(0,r)}, 0) \\ &= 0. \end{aligned}$$

Therefore, we have

$$\begin{aligned} &\Sigma_{j=1}^{N-1}[deg(I - \alpha_{j2}S - T(\alpha_{j2}, \cdot), B(0,r), 0) \\ &-deg(I - \alpha_{j1}S - T(\alpha_{j1}, \cdot), B(0,r), 0) = 0. \end{aligned}$$

Since the degree has a jump at α_0 and the jumps occur only at eigenvalues of odd algebraic multiplicity, it follows that the degree has an even number of jumps. Consequently, C contains an odd number of trivial zeros $(\alpha_i, 0) \neq (\alpha_0, 0)$.

2.5 Applications to ODEs and PDEs

In this section, we give some applications of the Leray Schauder theory to existence of solutions for ordinary differential equations and partial differential equations.

Theorem 2.5.1. (Peano's Theorem) Let $f : R \times B(x_0, r) \subset R^n \to R^n$ be a continuous function. Then there exists $t_0 > 0$ such that the following equation:

$$\begin{cases} x'(t) = f(t, x(t)), & t \in (0, t_0), \\ x(0) = x_0 \end{cases} \tag{E 2.5.1}$$

has a solution.

Proof. Since f is continuous, there exist $t_1 > 0$, $r_1 < r$, and $M > 0$ such that

$$|f(t,x)| \leq M, \quad t \in [-t_1, t_1], \quad |x - x_0| \leq r_1,$$

where $|\cdot|$ is the norm in R^n. Take $t_0 \in (0, t_1)$ such that $Mr_0 < r_1$. We set

$$E = C([0, t_0], R^n) = \{x(t) : [0, t_0] \to R^n \text{ is continuous}\}$$

with the norm $|x(\cdot)| = \max_{t\in[0,t_0]} |x(t)|$. Then $C([0,t_0], R^n)$ is a Banach space. Put

$$K = \{x(\cdot) \in E : x(0) = x_0, |x(t) - x_0| \le r_1, t \in [0,t_0]\}.$$

Then it is easy to see that K is a bounded closed convex subset.

Now, we define a mapping $T : E \to E$ by

$$Tx(t) = x_0 + \int_0^t f(s, x(s))ds \quad \text{for all } x(\cdot) \in E.$$

It is easy to check that T is continuous. Moreover, $T : K \to K$ is a mapping and $\overline{TK}$ is compact. Thus, by Schauder's fixed point theorem, T has a fixed point $x(\cdot)$ in K, i.e., $x(t) = x_0 + \int_0^t f(s, x(s))ds$. Therefore,

$$x'(t) = f(t, x(t)) \quad \text{for all } t \in (0, t_0).$$

Thus the problem (E 2.5.1) has a solution. This completes the proof.

In the following, let $\Omega \subset R^N$ be an open bounded subset with smooth boundary and $a_i, b : \overline{\Omega}\times R\times R^N \to R$, $i = 1, 2, \cdots, N$, be continuous functions such that

(1) $\frac{\partial a_i}{\partial \eta_j}\xi_i\xi_j \ge |\xi|^2$ for all $(x, z, \xi) \in \Omega \times R \times R^n$;

(2) $|a_i(x, z, 0) \le g(z)$, $i = 1, 2, \cdots, N$, where $g(\cdot) \in L^q(\Omega)$, and $q > N$;

(3) $(1 + |\xi|^2)|\frac{\partial a_i}{\partial \xi_j}| + (1 + |\xi|)(|\frac{\partial a_i}{\partial z}| + |a_i|) + |\frac{\partial a_i}{\partial x_j}| + |b| \le \mu(|z|)(1 + |\xi|^2)$ for $i, j = 1, 2 \cdots, N$, where $\mu : [0, +\infty) \to [0, +\infty)$ is a increasing function;

(4) $-b(x, z, \xi) signz \le L(|\xi| + f(x))$ for all $(x, z, \xi) \in \Omega \times R \times R^N$, where $L > 0$ is a constant.

Consider the following Dirichlet problem:

$$\begin{cases} -D_i a_i(x, u(x), Du(x)) + b(x, u(x), Du(x)) = 0, & x \in \Omega, \\ u(x) = \phi(x), & x \in \partial\Omega. \end{cases} \qquad (E\ 2.5.2)$$

Theorem 2.5.2. Suppose that $\partial\Omega$ is smooth and (1)-(4) hold and $a_i \in C^{1,\alpha}(\overline{\Omega} \times R \times R^N)$, $b \in C^{0,\alpha}(\overline{\Omega} \times R \times R^N)$, $\phi \in C^{2,\alpha}(\overline{\Omega})$, where $\alpha \in (0, 1)$ is a constant. Then the problem (E 2.5.2) has a solution $u \in C^{2,\alpha}(\overline{\Omega})$.

Proof. For each $v \in C^{1,\alpha}(\overline{\Omega})$ and $t \in [0, 1]$, consider the following linear Dirichlet problem:

$$\begin{cases} -[t\frac{\partial a_i}{\partial \eta_j}(x, v(x), Dv(x))D_{ij}u(x) + (1 - t)\Delta u(x)] & \\ +t[\frac{\partial a_i}{\partial x_i} + \frac{\partial a_i}{\partial z_i}\eta_i + b]_{(x,z,\eta)=(x,v(x),Dv(x))} = 0, & x \in \Omega, \\ u(x) = t\phi(x), & x \in \partial\Omega. \end{cases} \qquad (E\ 2.5.3)$$

Since the coefficients belong to $C^{\alpha}(\overline{\Omega})$, it is well known that $(E\ 2.5.3)$ has a unique solution $u \in C^{2,\alpha}(\overline{\Omega})$ (see [132]). We define a mapping $T : [0,1] \times C^{1,\alpha}(\overline{\Omega}) \to C^{1,\alpha}(\overline{\Omega})$ by

$$T(t,v) = u \quad \text{for all } (t,v) \in [0,1] \times C^{1,\alpha}(\overline{\Omega}).$$

It is easy to see that T is continuous compact. Also, by prior estimation, there exist constants $M > 0$, $0 < \gamma < 1$ such that $|u|_{1,\gamma;\Omega} \leq M$ (see [135]). Therefore, $u \in C^{\alpha\gamma}(\overline{\Omega})$ and there exist a constant C which does not depend on t and u such that

$$|u|_{2,\alpha\gamma} \leq C.$$

Thus $|u|_{1,\alpha} \leq M$. Therefore, $T(\cdot, 1)$ has a fixed point u, i.e., $T(1,u) = u$, which is a solution of $(E\ 2.5.2)$ and $u \in C^{2,\alpha}(\overline{\Omega})$.

In the following, suppose that H is a real Hilbert space, $A : D(A) \subseteq H \to H$ is a linear self-adjoint operator and $F(t,u) : R \times H \to H$ is a nonlinear mapping. Consider the anti-periodic problem:

$$\begin{cases} u' + Au(t) + \partial Gu(t) + F(t,u(t)) = 0, & a.e.\, t \in R, \\ u(t) = -u(t+T), & t \in R. \end{cases} \qquad (E\ 2.5.4)$$

Definition 2.5.3. A function $u(\cdot)$ is called a weak anti-periodic solution of $(E\ 2.5.4)$ if $u(t+T) = -u(t)$ for $t \in R$, $\int_0^T |u'(t)|^2 dt < \infty$ and

$$u' + Au(t) + \partial Gu(t) + F(t,u(t)) = 0 \quad \text{for almost all } t \in R.$$

Lemma 2.5.4. If $u, u' \in L^2(0,T;H)$ and $u(t+T) = -u(t)$ for all $t \in R$, then

$$|u|_\infty \leq \frac{\sqrt{T}}{2} \Big(\int_0^T |u'(s)|^2 ds \Big)^{\frac{1}{2}}.$$

Proof. Since $u(t) = u(0) + \int_0^t u'(s)ds$ and $u(t) = u(T) - \int_t^T u'(s)ds$, we have

$$u(t) = \frac{1}{2}[\int_0^t u'(s)ds - \int_t^T u'(s)ds].$$

Thus the conclusion follows.

Lemma 2.5.5. Let H be a real separable Hilbert space and $A : D(A) \subseteq H \to H$ be a linear densely defined closed self-adjoint operator that only has a point spectrum, i.e., eigenvalues. Suppose that $f : R \to H$ is a T-anti-periodic function, i.e., $f(T+t) = -f(t)$ for $t \in R$, and $f(\cdot) \in L^2(0,T;H)$. Then the following problem:

$$\begin{cases} u' + Au(t) + f(t) = 0, & a.e. t \in R, \\ u(t+T) = -u(t), & t \in R, \end{cases} \qquad (E\ 2.5.5)$$

has a unique weak solution.

Proof. Since $\sigma(A)$ only has point spectrum and H is separable, A has a countable family of eigenvalues $\{\lambda_i\}_{i=1}^{\infty}$. Assume that $\{e_i : i = 1, 2, \cdots\}$ is the orthogonal family of eigenvectors associated with the eigenvalue λ_i satisfying $|e_i| = 1$ for each i, $i = 1, 2 \cdots$. A is densely defined and we have $span\{e_i : i = 1, 2, \cdots\} = H$, so $f(t) = \Sigma_{i=1}^{\infty} f_i(t)e_i$, where $f_i : R \to R$ satisfies $\Sigma_{i=1}^{\infty} \int_0^T f_i^2(t)dt < \infty$. It is obvious that $f_i(t+T) = -f_i(t)$ for $t \in R$.

Now, we consider the one dimensional evolution equation:

$$\begin{cases} u_i'(t) + \lambda_i u_i(t) + f_i(t) = 0, & t \in R, \\ u_i(t+T) = -u_i(t), & t \in R, \end{cases} \tag{E 2.5.6}$$

for each $i = 1, 2, \cdots$. By Corollary 1.2 of [148] or by a direct computation, the problem (E 2.5.6) has a unique solution $u_i(t)$.

Now, multiply (E 2.5.6) by $u_i'(t)$ and integrate over $(0, T)$ to get

$$\int_0^T |u_i'(t)|^2 dt + \int f_i(t) u_i'(t) dt = 0.$$

Therefore, we have

$$\int_0^T |u_i'(t)|^2 dt \leq \int_0^T |f_i(t)|^2 dt. \tag{2.5.1}$$

By Lemma 2.5.4, we get

$$|u_i|_\infty^2 \leq \frac{T^2}{4} \int_0^T |f_i(t)|^2 dt. \tag{2.5.2}$$

Put $u(t) = \Sigma_{i=1}^{\infty} u_i(t)e_i$. Then it follows from (2.5.2) that u is well defined. By (2.5.1), we know that $u'(t) = \Sigma_{i=1}^{\infty} u_i'(t)e_i$ belongs to $L^2(0, T; H)$. Therefore, $\Sigma_{i=1}^{\infty} \lambda_i u_i(t)e_i$ belongs to $L^2(0, T; H)$. Since A is closed, $u(t) \in D(A)$ for almost all $t \in R$ and

$$Au(t) = \Sigma_{i=1}^{\infty} \lambda_i u_i(t) e_i \quad \text{for almost all } t \in R.$$

In view of (E 2.5.6), we know that u is a weak solution of the problem (E 2.5.5).

If u and v are two weak anti-periodic solutions of (E 2.5.5), then

$$\int_0^T [|u'(t) - v'(t)|^2 + (Au(t) - Av(t), u'(t) - v'(t))]dt = 0.$$

However, $\int_0^T (Au(t) - Av(t), u'(t) - v'(t))dt = 0$, so the uniqueness is obvious. This completes the proof.

Theorem 2.5.6. Let H be a real separable Hilbert space, $A : D(A) \subseteq H \to H$ be a linear densely defined closed self-adjoint operator that only has a point spectrum and $G : H \to R$ be a even continuous differentiable function such that the gradient ∂G is continuous and bounded, i.e., maps bounded sets of H to bounded sets in H. Suppose that $F : R \times H \to H$ is a continuous function and the following conditions are satisfied:

(1) $D(A)$ is compactly embedded into H.

(2) $F(t+T, -u) = -F(t,u)$ for all $(t,u) \in R \times H$.

(3) $|F(t,u)| \leq f(t)$, a. e. $t \in R$, where $f(\cdot) \in L^2(0,T;R)$ is non-negative.

Then the following equation:

$$\begin{cases} u' + Au(t) + \partial Gu(t) + F(t,u) = 0, & a.e.\, t \in R, \\ u(t+T) = -u(t), & t \in R, \end{cases} \tag{E 2.5.7}$$

has a weak solution.

Proof. Let

$$W_a = \{u : R \to H \text{ is continuous}, u(t+T) = -u(t)\},$$

$$W_a^{1,2} = \{u \in W_a : \int_0^T |u'(t)|^2 dt < \infty)\}.$$

For each $v(\cdot) \in W_a$, we consider the following equation:

$$\begin{cases} u' + Au(t) + \partial Gv(t) + F(t,v(t)) = 0, & a.e.\, t \in R, \\ u(t) = -u(t+T), & t \in R. \end{cases} \tag{E 2.5.8}$$

Since A only has a point spectrum, Lemma 2.5.5 implies that the problem (E 2.5.8) has a unique solution $u \in W_a^{1,2}$. Now, we define a mapping $K : W_a \to W_a$ as follows: For each $v \in W_a$, Kv is the unique solution of (E 2.5.8). Next, we prove that K is continuous. Suppose that $v_n \to v_0$ in W_a. Then, by Lemma 2.5.4, $|v_n - v|_\infty \to 0$ as $n \to \infty$. Now, ∂G and F are continuous functions, so

$$|\partial Gv_n(\cdot) - \partial Gv_0(\cdot)|_\infty \to 0, \quad |F(\cdot, v_n(\cdot)) - F(\cdot, v_0(\cdot))|_\infty \to 0$$

as $n \to \infty$. Also, we have

$$\begin{aligned} &(Kv_n(t))' - (Kv_0(t))' + A(Kv_n(t) - Kv_0(t)) \\ &\quad + (\partial GKv_n(t) - \partial GKv_0(t)) + F(t, v_n(t)) - F(t, v_0(t)) \\ &= 0 \quad \text{for almost all } t \in R. \end{aligned} \tag{2.5.3}$$

Multiply both sides of (2.5.3) by $(Kv_n(t) - Kv_0(t))'$ and integrate over $(0, T)$, and we have

$$\begin{aligned}&\int_0^T |(Kv_n(t) - Kv_0(t))'|^2 dt \\ &\quad + \int_0^T \langle A(Kv_n(t)) - Kv_0(t)), (Kv_n(t) - Kv_0(t))'\rangle dt \\ &\quad + \int_0^T \langle \partial G Kv_n(t) - \partial G Kv_0(t), (Kv_n(t) - Kv_0(t))'\rangle dt \\ &\quad + \int_0^T \langle F(t, v_n(t)) - F(t, v_0(t)), (Kv_n(t) - Kv_0(t))'\rangle dt = 0.\end{aligned}$$

Thus we have

$$\begin{aligned}&\Big(\int_0^T |(Kv_n(t) - Kv_0(t))'|^2 dt\Big)^{\frac{1}{2}} \\ &\leq \sqrt{T}|\partial G v_n(\cdot) - \partial G v_0(\cdot)|_\infty + \sqrt{T}|F(\cdot, v_n(\cdot)) - F(\cdot, v_0(\cdot))|_\infty \\ &\to 0 \quad (n \to \infty).\end{aligned}$$

Therefore, it follows that $Kv_n(\cdot) \to Kv_0(\cdot)$ in W_a. For each $v \in W_a$, again, by $(E\ 2.5.8)$, we get

$$\int_0^T ((Kv(t))')^2 dt + \int_0^T \langle \partial G v(t), (Kv(t))'\rangle dt + \int_0^T \langle F(t, v(t)), (Kv(t))'\rangle dt = 0.$$

This and assumption (3) of Theorem 2.5.6 imply that

$$\Big(\int_0^T |(Kv(t))'|^2 dt\Big)^{\frac{1}{2}} \leq \Big(\int_0^T |\partial G v(t)|^2 dt\Big)^{\frac{1}{2}} + \Big(\int_0^T f^2(t) dt\Big)^{\frac{1}{2}}. \qquad (2.5.4)$$

By (2.5.4) and the boundedness of ∂G, we know that K maps bounded sets of W_a to bounded sets in W_a The compact embedding of $D(A)$ into H implies that K is a compact mapping. In view of Lemma 2.5.4, we may simply take $|u'|_{L^2}$ as the norm of u in W_a.

Now, we prove that $Kv \neq \lambda v$ for $v \in W_a$ with $|v'|_{L^2} > (\int_0^T f^2(t)dt)^{\frac{1}{2}}$ and $\lambda > 1$. In fact, if this is not true, then there exist $\lambda_0 > 1$ and $v_0 \in W_a$ with $|v_0'|_{L^2} > (\int_0^T f^2(t)dt)^{\frac{1}{2}}$ such that $Kv_0 = \lambda_0 v_0$, i.e.,

$$\lambda_0 v_0'(t) + \lambda_0 A v_0(t) + \partial G v_0(t) + F(t, v_0(t)) = 0. \qquad (2.5.5)$$

Multiply both sides of (2.5.5) by $v_0'(t)$ and integrate over $[0, T]$, we obtain

$$\int_0^T \langle \lambda_0 v_0'(t) + \lambda_0 A v_0(t) + \partial G v_0(t) + F(t, v_0(t)), v_0'(t)\rangle dt = 0.$$

However, $\int_0^T \langle Av_0(t), v_0'(t)\rangle dt = 0$ and $\int_0^T \partial\langle Gv_0(t), v_0'(t)\rangle dt = 0$ and so we have

$$\int_0^T \langle \lambda_0 v_0'(t) + F(t, v_0(t)), v_0'(t)\rangle dt = 0.$$

Therefore, it follows that

$$\Big(\int_0^T \lambda_0 |v_0'(t)|^2 dt\Big)^{\frac{1}{2}} \le \Big(\int_0^T f^2(t)dt\Big)^{\frac{1}{2}},$$

which is a contradiction. Now, if we take $r_0 > (\int_0^T f^2(t)dt)^{\frac{1}{2}}$, then, by the above arguments and the homotopy invariance property of the Leray Schauder degree, we know that

$$deg(I - K, B(0, r_0), 0) = deg(I, B(0, r_0), 0) = 1,$$

where $B(0, r_0)$ is the open ball centered at 0 with radius r_0 in W_a. Thus K has a fixed point in $B(0, r_0)$, i.e., there exists $v \in W_a$ such that $Kv = v$. Hence the problem (E 2.5.7) has a solution. This completes the proof.

From Theorem 2.5.6, we have the following:

Corollary 2.5.7. Let H be a real separable Hilbert space, $A : D(A) \subseteq H \to H$ be a linear densely defined closed self-adjoint operator that only has a point spectrum, $0 \notin \sigma(A)$ and $G : H \to R$ be a even continuous differentiable function such that the gradient ∂G is Lipschitz. Suppose that $F : R \times H \to H$ is a continuous function and the conditions (1)-(3) in Theorem 2.5.6 are satisfied. Then the problem (E 2.5.7) has a solution.

Example 2.5.8. Consider the anti-periodic solution problem:

$$\begin{cases} u_1'(t) = 2u_1(t) - \alpha u_2(t) + \frac{2}{1+u_2^2(t)} \sin t, & t \in R, \\ u_2'(t) = -\alpha u_1(t) - 5u_2(t) + \frac{1}{1+u_1^4(t)} \sin^3 t, & t \in R, \\ u_1(t+\pi) = -u_1(t), u_2(t+\pi) = -u_2(t), & t \in R, \end{cases} \qquad (E\ 2.5.9)$$

where $\alpha \in R$ is a constant. Set

$$A = \begin{pmatrix} 2 & -\alpha \\ -\alpha & -5 \end{pmatrix}, \quad u = \begin{pmatrix} u_1 \\ u_2 \end{pmatrix}, \quad F(t, u) = \begin{pmatrix} \frac{2}{1+u_2^2(t)} \sin t \\ \frac{1}{1+u_1^4(t)} \sin^3 t \end{pmatrix}.$$

Then A is a linear self-adjoint operator on R^2 and $F(t + \pi, -u) = -F(t, u)$ for all $(t, u) \in R \times R^2$. It is obvious that $|F(t, u)| \le 3$ for $(t, u) \in R \times R^2$. Now, the problem (E 2.5.9) is equivalent to the equation:

$$\begin{cases} u'(t) = Au(t) + F(t, u(t)), & t \in R, \\ u(t + \pi) = -u(t), & t \in R. \end{cases} \qquad (E\ 2.5.10)$$

From Theorem 2.5.6, we know that the problem (E 2.5.10) has a solution, so (E 2.5.9) has a solution.

Remark. The degree theory in this chapter can be established in locally convex spaces (see [197]) or admissible topological vector spaces (see [170], [235], [307]).

2.6 Exercises

1. Let $f(x, y) : R \times R \to R$ be a continuous function and $T : C[a, b] \to C[a, b]$ be defined by $Tx(\cdot)(s) = \int_a^s f(t, x(t))dt$ for all $x(\cdot) \in C[a, b]$ and $s \in [a, b]$. Show that T is continuous and compact.

2. Let $T : C[0, \pi] \to C[0, \pi]$ be defined by

$$Tx(t) = \frac{2}{\pi} \int_0^\pi [\sin t \sin s + c \sin 2t \sin 2s][2x(s) + x^3(s)]ds$$

for all $x(\cdot) \in C[0, \pi]$. Compute $T'(x)$ and the eigenvalues of $T'(0)$.

3. Let $T : c_0 \to c_0$ be defined by $T(x_1, x_2, \cdots) = (x_1^2, x_2^2 \cdots)$. Show that $T'(x)$ is compact for all $x \in c_0$, but T is not compact.

4. Let E be a infinite dimensional Banach space. Show that the unit sphere of E is not compact.

5. Let E be a real Banach space and Ω be an open subset of E. Let $T : \Omega \to E$ be a continuous compact mapping and $I - T$ is locally one to one. Show that $I - T$ is an open mapping.

6. Let E be a real Banach space, $T : \overline{B(0, R)} \to E$ be a continuous compact mapping and $L : X \to X$ be a compact linear operator. Suppose that $\|Tx - Lx\| < \|x - Lx\|$ for all $x \in \partial B(0, R)$. Show that $deg(I - T, B(0, R), 0)$ is odd.

7. Let E be a real Banach space, $\Omega \subset E$ be an open bounded symmetric subset with $0 \in \Omega$ and $T : \overline{\Omega} \to E$ be a continuous compact mapping such that $0 \notin (I - T)(\partial\Omega)$ and $(I - T)(-x) \neq t(I - T)(x)$ on $\partial\Omega$ for all $t \geq 1$. Show that $deg(I - T, \Omega, 0)$ is odd.

8. Let E be a real Banach space and $T : E \to E$ be a continuous compact mapping such that $\lim_{\|x\| \to \infty} \frac{\|Tx\|}{\|x\|} = 0$. Show $(I - T)(E) = E$

9. Let E be a real Banach space and $A, B \subset E$ be two closed bounded subsets. Suppose that there exists a continuous compact mapping T :

$A \to E$ such that $I - T : A \to B$ is a homeomorphism. Show that $E \setminus A$ and $E \setminus B$ have the same number of components.

10. Let E be a locally convex space and $\Omega \subset E$ be an open subset. Suppose that $T : \overline{\Omega} \to E$ is continuous such that $\overline{T(\overline{\Omega})}$ is compact in E and $0 \notin (I - T)(\partial\Omega)$. Show that there is a topological degree $deg(I - T, \Omega, 0)$.

11. Is a Hausdorff continuous mapping continuous?

12. Let E be a real Banach space, $\Omega \subset E$ be an open bounded symmetric subset with $0 \in \Omega$ and $T : \overline{\Omega} \to 2^E$ be an upper semicontinuous compact mapping with closed convex values such that $0 \notin (I - T)(\partial\Omega)$ and $T(-x) = -Tx$ for all $x \in \partial\Omega$. Show that $deg(I - T, \Omega, 0)$ is odd.

13. Let $G : R^n \to R$ be an even continuous differentiable function and $f : R \to R^n$ be a continuous function such that $f(t + T) = -f(t)$ for all $t \in R$. Show that the following equation

$$\begin{cases} x'(t) = \partial G x(t) + f(t), & t \in R, \\ x(t + T) = -x(t), & t \in R, \end{cases}$$

has a solution.

14. Let $g \in C^1([0, a])$, $h \in C^1([0, b])$ with $g(0) = h(0)$ and $f : [0, a] \times [0, b] \times R^2 \to R$ be continuous such that $|f(x, y, z, 0)| \le M(1 + |z|)$ and $|f(x, y, z, u) - f(x, y, z, v)| \le L|u - v|$, where a, b, L, M are positive constants. Show that the following equation

$$\begin{cases} \frac{\partial u}{\partial x \partial y} = f(x, y, u(x, y), \frac{\partial u}{\partial x}), & (x, y) \in [0, a] \times [0, b], \\ u(x, 0) = g(x), & x \in [0, a], \\ u(0, y) = h(y), & y \in [0, b], \end{cases}$$

has a solution $u \in C^1([0, a] \times [0, b])$.

Chapter 3

DEGREE THEORY FOR SET CONTRACTIVE MAPS

The Leray Schauder degree theory is very useful in solving an operator equation of the type $(I - T)x = y$, where T is compact. In many applications T is not compact, so one may ask it is possible to give an analogue of the Leray Schauder theory in the noncompact case. In 1936, Leray [184] constructed an example to show that it is impossible to define a degree theory for mappings with only a continuity condition. So a very natural question which arises is the following:

For what kind of mappings in infinite dimensional spaces can we establish a degree theory ?

Browder, Nussbaum, Sadovski, Vath, etc., showed that it is possible to define a complete analogue of the Leray Schauder theory for condensing type mappings T.

In this Chapter, we will introduce the degree theory for k-set contraction mappings and condensing mappings. This chapter consists of three sections.

In Section 3.1, we define measures of non-compactness and present some properties (see propositions 3.1.5, 3.1.7 and theorems 3.1.14, 3.1.15). Also, countably condensing maps, etc., are defined here and, in particular, a fixed point theorem for countably condensing self-mappings is presented in Corollary 3.1.18.

Section 3.2 presents a degree theory for countably condensing mappings and the theory is based on the use of retractions and the Leray Schauder degree. Again, various properties and consequences are presented.

In Section 3.3, we use the degree of Section 3.2 to discuss the initial and anti-periodic ordinary differential equations in Banach spaces.

3.1 Measure of Noncompactness and Set Contractions

Let (X, d) be a metric space and $A \subset X$ be a subset. We call $diam(A) = sup_{x,y \in A} d(x, y)$ the diameter of A. If $diam(A) < +\infty$, then we call A

bounded. For two bounded sets A, B, the Hausdorff metric H is defined by

$$H(A,B) = \max\{\sup_{x\in A} d(x,B), \sup_{y\in B} d(y,A)\}.$$

Let $B(X)$ be the collection of all bounded subsets of X.

Proposition 3.1.1. If $A \subset B$, then $diam(A) \le diam(B)$ and $diam(\overline{A}) = diam(A)$.

Proposition 3.1.2. Let X be a Banach space and $A, B \subset X$. Then we have the following:

(1) $diam(\lambda B) = |\lambda| diam(B)$;

(2) $diam(x + B) = diam(B)$;

(3) $diam(A + B) \le diam(A) + diam(B)$;

(4) $diam(conv(A)) = diam(A)$.

Proof. (1)-(3) are obvious. To see (4), let $x, y \in conv(A)$. There exist $s_i \in (0,1)$, $x_i \in X$, $i = 1, 2, \cdots, k$, $t_i \in (0,1)$, $y_i \in A$, $i = 1, 2, \cdots, m$, such that $x = \Sigma_{i=1}^k s_i x_i$ and $y = \Sigma_{i=1}^m s_i y_i$. Now, we have

$$\begin{aligned}\|x - y\| &= \|\Sigma_{i=1}^k s_i x_i - \Sigma_{i=1}^m s_i y_i\| \\ &= \Sigma_{i=1}^k \Sigma_{j=1}^m s_i t_j x_i - \Sigma_{i=1}^k \Sigma_{j=1}^m s_i t_j y_j\| \\ &\le \Sigma_{i=1}^k \Sigma_{j=1}^m s_i t_j \|x_i - y_j\| \\ &\le \Sigma_{i=1}^k \Sigma_{j=1}^m s_i t_j diam(A).\end{aligned}$$

Thus $diam(conv(A)) \le diam(A)$, and the converse is obvious. Therefore, the conclusion is true. This completes the proof.

Proposition 3.1.3. Let (X, d) be a metric space. Then $(B(X), H)$ is a metric space.

Proof. Obviously, $H(A, B) \ge 0$ for any $A, B \in B(X)$, $H(A, B) = 0$ if only if $A = B$ and $H(A, B) = H(B, A)$.

For $A, B, C \in B(X)$, we have

$$d(x, B) \le d(x, z) + d(z, B), \quad d(y, A) \le d(y, z) + d(z, A)$$

for all $z \in C$, $x \in A$ and $y \in B$. Thus

$$d(x, B) \le \inf_{z\in C} d(x, z) + \sup_{z\in C} d(z, B),$$

$$d(y, A) \le \inf_{z\in C} d(y, z) + \sup_{z\in C} d(z, A)$$

and so

$$
\begin{aligned}
&H(A,B)\\
&\leq \max\{\sup_{x\in A} d(x,C) + \sup_{z\in C} d(z,B), \sup_{y\in B} d(y,C) + sup_{z\in C} d(z,A)\}\\
&\leq \max\{\sup_{x\in A} d(x,C), sup_{z\in C} d(z,A)\} + \max\{\sup_{z\in C} d(z,B), \sup_{y\in B} d(y,C)\}\\
&= H(A,C) + H(C,B).
\end{aligned}
$$

Thus $(B(X),H)$ is a metric space. This completes the proof.

Definition 3.1.4. Let (X,d) be a metric space, $\mathcal{B}$ be the collection of all bounded subsets of X and $A,B \in \mathcal{B}$. A function $\alpha : \beta \to [\prime, +\infty)$ defined by

$$
\begin{aligned}
\alpha(A) = \inf\{\delta > 0 : A &\text{ is covered by finitely}\\
&\text{many sets each having diameter less than } \delta\}
\end{aligned}
$$

is called the (Kuratowski) measure of noncompactness. If we replace $\alpha(A)$ by

$$\beta(A) = \inf\{\delta > 0 : A \text{ is covered by finitely many balls with radius } \delta\},$$

then we call $\beta(A)$ as the ball (Hausdorff) measure of noncompactness.

The relation between α and β is given by the following inequality.

Proposition 3.1.5. $\beta(A) \leq \alpha(A) \leq 2\beta(A)$ for all $A \in \mathcal{B}$.

Proof. For any $\delta > \alpha(A)$, there exist finitely many sets $A_1, A_2, \cdots, A_k$ such that $A \subset \cup_{i=1}^k A_i$ and $diam(A_i) < \delta$ for $i = 1, 2, \cdots, k$. Choose $x_i \in A_i$. Then $B_\delta(x_i) \supset A_i$, so we have $A \subset \cup_{i=1}^k B_\delta(x_i)$ and thus $\beta(A) \leq \delta$. By letting $\delta \to \alpha(A)$, we get the first inequality.

For the second inequality, if $\delta > \beta(A)$, then there exist finitely many balls $B_\delta(y_1), B_\delta(y_2), \cdots, B_\delta(y_m)$ such that $A \subset \cup_{i=1}^m B_\delta(y_i)$. It is obvious that

$$diam(B_\delta(y_i)) = 2\delta,$$

so we have $\alpha(A) \leq 2\delta$ and, by letting $\delta \to \beta(A)$, we deduce the desired inequality. This completes the proof.

Proposition 3.1.6. Let X be a metric space and $\mathcal{B}$ be the collection of all bounded subsets of X. Let ϕ be the Kuratowski measure or Hausdorff measure of noncompactness and $A, B \in \mathcal{B}$. Then the following properties hold:

(1) $\phi(A) = 0$ if and only if A is relatively compact;

(2) $\phi(\overline{A}) = \phi(A)$;

(3) If $A \subset B$, then $\phi(A) \leq \phi(B)$;

(4) $\phi(A \cup B) = \max\{\phi(A), \phi(B)\}$;

(5) $\phi(A \cap B) \leq \min\{\phi(A), \phi(B)\}$;

(6) $|\phi(A) - \phi(B)| \leq 2H(A, B)$;

(7) If X is a Banach space, then

$$\phi(\lambda A) = |\lambda|\phi(A), \lambda \in R, \quad \phi(A+B) \leq \phi(A)+\phi(B), \quad \phi(conv(A)) = \phi(A).$$

Proof. In (1)–(5), $\phi(\lambda A) = |\lambda|\phi(A)$ for any $\lambda \in R$ and $\phi(A + B) \leq \phi(A) + \phi(B)$ follow easily from the definition.

We only need to prove (6) and $\phi(conv(A)) = \phi(A)$. Assume that $\phi = \alpha$. The proof is the same for $\phi = \beta$.

For any $\epsilon > 0$, there exists a finite cover $\{A_1, A_2, \cdots, A_k\}$ of A with $diam(A_i) \leq \alpha(A) + \epsilon$ for $i = 1, 2, \cdots, k$. Set

$$\eta = H(A, B) + \epsilon, \quad B_i = \{y \in B : \text{ there exists } x \in A_i, d(x, y) < \eta\}$$

for $i = 1, 2, \cdots, k$. Since $H(A, B) < \eta$, we have $B \subset \cup_{i=1}^k B_i$. Obviously,

$$diam(B_i) \leq 2\eta + diam(A_i) < 2H(A, B) + \alpha(A) + 3\epsilon$$

for $i = 1, 2, \cdots$ and thus $\alpha(B) \leq 2H(A, B) + \alpha(A)$.

Similarly, we have $\alpha(A) \leq 2H(A, B) + \alpha(B)$. Therefore, we have

$$|\alpha(A) - \alpha(B)| \leq 2H(A, B).$$

For (7): Obviously, $\alpha(A) \leq \alpha(conv(A))$. For any $\epsilon > 0$, there exists a finite cover $\{A_1, A_2, \cdots, A_k\}$ of A with $diam(A_i) < \alpha(A) + \epsilon$ for $i = 1, 2, \cdots, k$. We may also assume that B_i is convex since $diam(conv(B_i)) = diam(B_i)$ for $i = 1, 2, \cdots, k$. Put

$$\Lambda = \{(\lambda_i, \lambda_2, \cdots, \lambda_k) : \lambda_i \geq 0, \ i = 1, 2, \cdots, k, \ \Sigma_{i=1}^k \lambda_i = 1\}$$

and $B(\lambda) = \Sigma_{i=1}^k \lambda_i B_i$ for each $\lambda = (\lambda_1, \lambda_2, \cdots, \lambda_k) \in \Lambda$. We have

$$\alpha(B(\lambda)) \leq \alpha(A) + \epsilon \quad \text{for all } \lambda \in \Lambda.$$

Now, we show that $\cup_{\lambda \in \Lambda} B(\lambda)$ is convex. For $\lambda = (\lambda_1, \lambda_2, \cdots, \lambda_k)$, $\mu = (\mu_1, \mu_2, \cdots, \mu_k) \in \Lambda$ and $x = \Sigma_{i=1}^k \lambda_i x_i \in B(\lambda)$, $y = \Sigma_{i=1}^k \mu_i y_i \in B(\mu)$, where $x_i, y_i \in B_i$ for $i = 1, 2, \cdots, k$, we have

$$tx + (1-t)y = \Sigma_{i=1}^k (t\lambda_i + (1-t)\mu_i)[\frac{t\lambda_i}{t\lambda_i + (1-t)\mu_i} x_i + \frac{(1-t)\mu_i}{t\lambda_i + (1-t)\mu_i} y_i].$$

Thus it follows that $\cup_{\lambda \in \Lambda} B(\lambda)$ is convex. Therefore, we have

$$conv(A) \subset conv(\cup_{i=1}^k B_i) \subset \cup_{\lambda \in \Lambda} B(\lambda).$$

Since Λ is compact, there exist finitely many $\lambda^1, \lambda^2, \cdots, \lambda^n \in \Lambda$ such that

$$\cup_{\lambda\in\Lambda} B(\lambda) \subset \cup_{i=1}^n B(\lambda^i) + \epsilon B_1(0).$$

Therefore, we have

$$conv(A) \subset \cup_{i=1}^n B(\lambda^i) + \epsilon B_1(0),$$

which implies that

$$\alpha(conv(A)) \leq \alpha(A) + 3\epsilon.$$

By letting $\epsilon \to 0^+$, we get $\alpha(conv(A)) \leq \alpha(A)$. Thus $\alpha(conv(A)) = \alpha(A))$. This completes the proof.

Proposition 3.1.7. Let X be a infinite dimensional Banach space and $B(0,1)$ be the unit ball. Then $\alpha(B(0,1)) = 2$.

Proof. It is obvious that $diam(B(0,1)) = 2$, so $\alpha(B(0,1)) \leq 2$.
If $\alpha(B(0,1)) < 2$, then there exists $A_1, A_2, \cdots, A_k$ such that $diam(A_i) < 2$ for $i = 1, 2\cdots, k$ and $B(0,1) \subset \cup_{i=1}^k A_i$. Take a k-dimensional subspace X_k of X and set $B_k(0,1) = B(0,1) \cap X_k$, $B_i = A_i \cap X_k$ for $i = 1, 2, \cdots, k$. Then we have

$$B_k(0,1) \subset \cup_{i=1}^k B_i, \quad diam(B_i) < 2, i = 1, 2, \cdots, k,$$

which is a contradiction to Theorem 1.2.17. Thus $\alpha(B(0,1)) = 2$. This completes the proof.

Proposition 3.1.8. Let X be a separable Banach space and β be the Hausdorff measure of noncompactness. Then there exists an increasing sequence of finite dimensional subspaces (X_n) with $X = \overline{\cup_{n=1}^\infty X_n}$ such that, for any bounded countable subset $\{x_n : 1 \leq n < \infty\}$,

$$\beta(\{x_n\}) = \lim_{n\to\infty} \limsup_{m\to\infty} d(x_m, X_n).$$

Proof. Since X is separable, there exists a countable subset $\{y_1, y_2, \cdots\}$ of X such that $\overline{\{y_1, y_2, \cdots\}} = X$. Put $X_n = span\{y_1, y_2, \cdots, y_n\}$ for $n = 1, 2, \cdots$. Then $X_1 \subset X_2 \subset \cdots$ and $X = \overline{\cup_{n=1}^\infty X_n}$. For any $\epsilon > 0$ and integer $n > 0$, set $r_n = \limsup_{m\to\infty} d(x_m, X_n)$. One can easily see that $r_1 \geq r_2 \geq \cdots$ since X_n is increasing. Choose an integer $L > 0$ such that $d(x_m, X_n) < r_n + \epsilon$ for all $m \geq L$.

Next, we define

$$Y = \{y \in X_n : \text{ there exists } m \geq L,\, d(x_m, X_n) = \|x_m - y\|\}.$$

Then it follows that $Y \cup \{x_1, x_2, \cdots, x_L\}$ is compact. Thus there exist finitely many $\{z_1, z_2, \cdots, z_k\}$ such that

$$Y \cup \{x_1, x_2, \cdots, x_L\} \subset \cup_{i=1}^k B(z_i, \epsilon).$$

Therefore, we have

$$\{x_1, x_2, \cdots\} \subset \{x_1, x_2, \cdots, x_L\} \cup B(Y, r_n + \epsilon) \subset \cup_{i=1}^{k} B(z_i, r_n + 2\epsilon).$$

Thus $\beta(\{x_n\}) \leq r_n + 2\epsilon$, i.e.,

$$\beta(\{x_n\}) \leq \inf\{r_n, n \geq 1\} = \lim_{n\to\infty} \limsup_{m\to\infty} d(x_m, X_n).$$

On the other hand, for $\epsilon > 0$, put $r = \beta(\{x_i : i \geq 1\})$ and there exists finitely many w_i, $1 \leq i \leq s$, such that $\{x_1, x_2, \cdots\} \subset \cup_{i=1}^{s} B(w_i, r + \epsilon)$. By the construction of X_n, we know that there exists an integer $K > 0$ such that $d(w_i, X_n) < \epsilon$ for $i = 1, 2, \cdots, s$ and $n > K$. Therefore, we have

$$\begin{aligned} d(x_m, X_n) &\leq \inf\{\|x_m - w_i\| : 1 \leq i \leq s\} + \sup\{d(w_i, X_n) : 1 \leq i \leq s\} \\ &\leq r + 2\epsilon \end{aligned}$$

for $m \geq 1$ and $n > K$. From this, we get

$$\lim_{n\to\infty} \limsup_{m\to\infty} d(x_m, X_n) \leq r.$$

Thus we have $\lim_{n\to\infty} \limsup_{m\to\infty} d(x_m, X_n) = \beta(\{x_i : i \geq 1\})$. This completes the proof.

Definition 3.1.9. Let X be a real normed space, $T : D \to X$ be a mapping and α be the measure of noncompactness.

(1) T is called a k-set contraction if $\alpha(TB) \leq k\alpha(B)$ for all bounded subsets $B \subset D$, where $k > 0$ is a constant;

(2) T is said to be condensing if $\alpha(TB) < \alpha(B)$ for all bounded subsets $B \subset D$ with $\alpha(B) > 0$.

Example 3.1.10. Let X be a real normed space and $T : X \to X$ be a linear bounded operator. Then L is a $\|L\|$-set contraction.

Example 3.1.11. Let X be a real normed space and $T : D \subset X \to X$ be a Lipschitz mapping with Lipschitz constant l. Then T is an l-set contraction.

Proposition 3.1.12. Let X be a real normed space, $\overline{B(0,1)}$ be the unit ball of X and $T : X \to \overline{B(0,1)}$ be defined by

$$Tx = \begin{cases} \frac{x}{\|x\|}, & \|x\| \geq 1, \\ x, & x \in B(0,1). \end{cases}$$

Then T is an 1-set contraction.

Proof. Let $A \subset X$ be a bounded set. It is obvious that $T(A) \subset conv(\{0\} \cup A)$, so we have

$$\alpha(T(A)) \leq \alpha(conv(\{0\} \cup A)) = \alpha(\{0\} \cup A) = \alpha(A).$$

Proposition 3.1.13. Let X be a infinite dimensional Banach space, $\phi : [0,1] \to [0,1]$ be a strictly decreasing and continuous function, $B(0,1)$ be the unit ball of X and $T : \overline{B(0,1)} \to \overline{B(0,1)}$ be defined by

$$Tx = \phi(|x|)x \quad \text{for all } x \in \overline{B(0,1)}.$$

Then $\alpha(T(B)) < \alpha(B)$ for $B \subset \overline{B(0,1)}$ with $\alpha(B) > 0$.

Proof. Let $B \subset \overline{B(0,1)}$ be such that $\alpha(B) = c > 0$. Take $r \in (0, \frac{c}{2})$ and define the sets $B_1 = B \cap \overline{B(0,r)}$, $B_2 = B \setminus \overline{B(0,r)}$. It is obvious that $T(B) = T(B_1) \cup T(B_2)$, so we have

$$\alpha(T(B)) = \alpha(T(B_1) \cup T(B_2)) \leq \max\{\alpha(T(B_1)), \alpha(T(B_2))\}.$$

Moreover, it follows that

$$\alpha(T(B_1)) \leq \alpha(conv(\{0\} \cup B_1)) = \alpha(B_1) \leq diam(B_1) \leq 2r < c$$

and

$$T(B_2) \subset conv(\{0\} \cup \phi(r)B),$$

so $\alpha(T(B_2)) \leq \alpha(\phi(r)B) < \alpha(B)$. Thus we have $\alpha(T(B)) < \alpha(B)$. This completes the proof.

Definition 3.1.14. Let X be a real Banach space, $T : D \to X$ be a mapping and α be the measure of noncompactness.

(1) T is called a countably k-set contraction if $\alpha(TB) \leq k\alpha(B)$ for all countably bounded subsets $B \subset D$, where $k > 0$ is a constant;

(2) T is said to be countably condensing if $\alpha(TB) < \alpha(B)$ for all countably bounded subsets $B \subset D$ with $\alpha(B) > 0$;

(3) $H(t,x) : [0,1] \times D \to X$ is said to be a homotopy of countably condensing mappings if $\alpha(H([0,1] \times B)) < \alpha(B)$ for all countably bounded subsets $B \subset D$ with $\alpha(B) > 0$.

One can easily see that a condensing mapping is a countably condensing mapping.

Theorem 3.1.15. Let E be a Banach space and $B \subset C([a,b], E)$ be a bounded equicontinuous subset. Then $\alpha(B(t))$ is continuous on $[a,b]$, where $B(t) = \{x(t) : x(\cdot) \in B\}$, and

$$\alpha(\{\int_a^b x(t)dt : x(\cdot) \in B\}) \leq \int_a^b \alpha(B(t))dt.$$

Proof. First, we prove that $\alpha(\{x(t) : x(\cdot) \in B\})$ is a continuous function on $[a,b]$. For any $\epsilon > 0$, since B is equicontinuous, there exists $\gamma > 0$ such that $\|x(t) - x(t')\| < \epsilon$ for all $t, t' \in [a,b]$ satisfying $|t - t'| < \gamma$. Thus we have

$$H(\{x(t) : x(\cdot) \in B\}, \{x(t') : x(\cdot) \in B\}) \leq \epsilon$$

for $t, t' \in [a, b]$ satisfying $|t - t'| < \gamma$. From this and (6) of Proposition 3.1.6, we infer that

$$|\alpha(\{x(t) : x(\cdot) \in B\}) - \alpha(\{x(t') : x(\cdot) \in B\})| \leq 2\epsilon$$

for $t, t' \in [a, b]$ satisfying $|t - t'| < \gamma$. Thus $\alpha(\{x(t) : x(\cdot) \in B\})$ is a continuous function on $[a, b]$.

For any division of $[a, b]$: $a = t_0 < t_1 < \cdots < t_n = b$, where $t_i = a + i\frac{b-a}{n}$, $i = 0, 1, \cdots, n$. For any $\epsilon > 0$, from the equicontinuity of B, there exists $N > 0$ such that, if $n > N$, then

$$\|x(t_i) - x(t)\| < \epsilon \quad \text{for all } x(\cdot) \in B,\ t \in [t_{i-1}, t_i]$$

for $i = 1, 2, \cdots, n$. Thus we have

$$\|\Sigma_{i=1}^n x(t_i)\frac{b-a}{n} - \int_a^b x(t)dt\| = \|\Sigma_{i=1}^n \int_{t_{i-1}}^{t_i} (x(t_i) - x(t))dt\| < \epsilon(b - a)$$

for all $n > N$. Therefore, we have

$$|\alpha(\{\Sigma_{i=1}^n x(t_i)\frac{b-a}{n} : x(\cdot) \in B\}) - \alpha(\{\int_a^b x(t)dt : x(\cdot) \in B\})| \leq 2\epsilon(b - a),$$

i.e.,

$$\lim_{n\to\infty} \alpha(\{\Sigma_{i=1}^n x(t_i)\frac{b-a}{n} : x(\cdot) \in B\}) = \alpha(\{\int_a^b x(t)dt : x(\cdot) \in B\}).$$

On the other hand, by (6) of Proposition 3.1.6, we have

$$\alpha(\{\Sigma_{i=1}^n x(t_i)\frac{b-a}{n} : x(\cdot) \in B\}) \leq \Sigma_{i=1}^n \alpha(\{x(t_i) : x(\cdot) \in B\})\frac{b-a}{n}.$$

Therefore, it follows that

$$\alpha(\{\int_a^b x(t)dt : x(\cdot) \in B\}) \leq \int_a^b \alpha(B(t))dt.$$

This completes the proof.

Theorem 3.1.16. Let E be a Banach space and $B \subset C([a, b], E)$ be a bounded equicontinuous subset, where $a, b \in R$. Then

$$\alpha(B) = \max_{t\in[a,b]} \alpha(\{x(t) : x(\cdot) \in B\}).$$

Proof. First, by Theorem 3.1.15, we know that $\alpha(\{x(t) : x(\cdot) \in B\})$ is a continuous function on $[a, b]$.

Next, for $\alpha(B) < \delta$, there exist $B_1, B_2, \cdots, B_m \subset C([a,b), E)$ such that $diam(B_i) \leq \delta$ and $B \subset \cup_{i=1}^m B_i$. Hence, for each $t \in [a,b]$, we have

$$\{x(t) : x(\cdot) \in B\} \subset \cup_{i=1}^m \{x(t) : x(\cdot) \in B_i\}.$$

Also, we have

$$diam(\{x(t) : x(\cdot) \in B_i\} = \sup_{x(\cdot), y(\cdot) \in B_i} \|x(t) - y(t)\| \leq diam(B_i) \leq \delta.$$

Therefore, $\alpha(\{x(t) : x(\cdot) \in B\}) \leq \delta$ for all $t \in [a,b]$ and thus

$$\max \alpha(\{x(t) : x(\cdot) \in B\}) \leq \alpha(B).$$

On the other hand, since B is equicontinuous, there exist finitely many $t_1, t_2, \cdots, t_n \in [a,b]$ such that

$$\{x(t) : x(\cdot) \in B\} \subset \cup_{i=1}^n \{x(t_i) : x(\cdot) \in B\} + B_\epsilon(0) \quad \text{for all} \quad t \in [a,b].$$

If $\delta > \max_{t \in [a,b]} \alpha(\{x(t) : x(\cdot) \in B\})$, we can find finitely many subsets $A_1, A_2, \cdots, A_s \subset E$ such that

$$diam(A_i) \leq \delta, \quad \cup_{i=1}^n \{x(t_i) : x(\cdot) \in B\} \subset \cup_{i=1}^s A_i.$$

Obviously, B is the union of finitely many sets $\{x(\cdot) \in B : x(t_i) \in A_{j_i}, i = 1, 2, \cdots, n\}$ and each of these sets has diameter less than $\delta + 2\epsilon$. Thus we have $\alpha(B) \leq \delta + 2\epsilon$ and we get the desired result. This completes the proof.

Proposition 3.1.17. Let E be a Banach space, $\Omega \subset E$ be an bounded subset and $T : \overline{\Omega} \to E$ be a countably condensing mapping. Put $F = \{x \in \overline{\Omega} \ Tx = x\}$. Then there exists a convex compact subset C such that

(1) $F \subseteq C$;

(2) if $x_0 \in \overline{conv(C \cup \{Tx_0\})}$, then $x_0 \in C$;

(3) $C = \overline{conv(T(C \cap \overline{\Omega}))}$.

Proof. Put

$$\mathcal{F} = \{K : F \subset K \subset E \text{ closed convex, } T(K \cap \overline{\Omega}) \subseteq K \text{ and (2) holds for } K\}.$$

Then $\mathcal{F}$ is nonempty since $\overline{conv(T\overline{\Omega})} \in \mathcal{F}$. We set $C = \cap_{K \in \mathcal{F}} K$. Obviously, C satisfies (1),(2), (3), and C is closed convex.

Now, we prove that C is compact. Assume that this is not true. Then there exists $C_1 = \{x_1, x_2, \cdots\} \subset C$ without a Cauchy subsequence. Note that $C = \overline{conv(T(C \cap \overline{\Omega}))}$. Thus there exists a countable subset $A_1 \in C \cap \overline{\Omega}$ such that $C_1 \subseteq \overline{conv(TA_1)}$. One can easily prove that $H_1 = \overline{conv(T(C_1 \cap \overline{\Omega}))}$ is separable and $H_1 \cap \overline{\Omega}$ is separable, so there exist countable subsets $B_1 \subset H_1$ and $D_1 \subset H_1 \cap \overline{\Omega}$ such that $\overline{B_1} = H_1$, $\overline{D_1} = H_1 \cap \overline{\Omega}$. Put $C_2 = C_1 \cup A_1 \cup B_1 \cup D_1$.

One can check that

$$C_1 \subset C_2, \tag{3.1.1}$$

$$\overline{conv(T(C_1 \cap \overline{\Omega}))} \subset \overline{C_2}, \tag{3.1.2}$$

$$\overline{conv(T(C_1 \cap \overline{\Omega})} \cap \overline{\Omega} \subset \overline{C_2 \cap \overline{\Omega}}. \tag{3.1.3}$$

In general, we define inductively a sequence (C_n) of countable subsets C_n of C satisfying

$$C_n \subset C_{n+1}, \tag{3.1.4}$$

$$\overline{conv(T(C_n \cap \overline{\Omega}))} \subset \overline{C_{n+1}}, \tag{3.1.5}$$

$$\overline{conv(T(C_n \cap \overline{\Omega})} \cap \overline{\Omega} \subset \overline{C_{n+1} \cap \overline{\Omega}}. \tag{3.1.6}$$

Finally, we put $L = \cup_{n=1}^{\infty} C_n$. Then

$$L \subseteq \overline{convT(L \cap \overline{\Omega}}.$$

Thus we have

$$\alpha(L) \leq \alpha(\overline{convT(L \cap \overline{\Omega}}) < \alpha(L \cap \overline{\Omega}),$$

which is a contradiction. Therefore, C is compact. This completes the proof.

Proposition 3.1.18. [291] Let E be a Banach space, $\Omega \subset E$ be a bounded subset and $T : \overline{\Omega} \to E$ be a countably condensing mapping. Set $C_1 = \overline{conv(T\overline{\Omega})}$, $C_{n+1} = \overline{conv(T(C_n \cap \overline{\Omega}))}$ for $n \geq 1$ and $C = \cap_{n=1}^{\infty} C_n$. Then C is convex and compact.

Proof. This is a special case of Lemma 7.2.1.

Corollary 3.1.19. Let $C \subset E$ be a nonempty bounded closed convex subset and let $T : C \to C$ be a continuous countably condensing mapping. Then T has a fixed point in C.

Proof. We first assume that T is k-set countably condensing for some $k \in [0, 1)$. Let $C_1 = \overline{convTC}$ and $C_{i+1} = \overline{convTC_i}$ for $i = 1, 2, \cdots$. By Proposition 3.1.18, $K = \cap_{i=1}^{\infty} C_i$ is convex and compact and also $T : K \to K$ is a mapping.

Now, we prove that K is non-empty. Take $x_0 \in C$, then $T^i x_0 \in C_i$ for $i \geq 1$. We have $\alpha(\{T^i x_0, i \geq n\}) \leq k^n \alpha(\{T_0^i, i \geq 0\})$ for $n \geq 1$. Obviously,

$$\alpha(\{T^i x_0, i \geq n\}) = \alpha(\{T^i x_0, i \geq 0\}),$$

so $\alpha(\{T^i x_0, i \geq 0\}) = 0$. Therefore, $(T^i x_0)_{i=1}^{\infty}$ has a subsequence which converges to a point in K and hence K is nonempty. Therefore, T has a fixed point.

Now, assume that T is countably condensing. Take $x_0 \in C$ and, for any $k \in (0,1)$, put $Sx = kTx + (1-k)x_0$ for all $x \in C$. Then S is k-set countably condensing, so S has a fixed point in C. Let $k_n \to 1$. Then there are $x_n \in C$ such that

$$k_n T x_n + (1 - k_n) x_0 = x_n,$$

which implies that $\alpha(\{Tx_n : n \geq 1\}) = \alpha(\{x_n : n \geq 1\})$. Since T is countably condensing, $(x_n)_{n=1}^{\infty}$ has a subsequence $x_{n_j} \to y \in C$. By the continuity of T, we have $Ty = y$. This completes the proof.

3.2 Degree Theory for Countably Condensing Mappings

In this section, we introduce a degree theory for condensing mappings.

Construction of the degree for a countably condensing mapping.

Let E be a Banach space and $\Omega \subset E$ be an open bounded subset. Let $T : \overline{\Omega} \to E$ be a continuous and countably condensing mapping and $0 \notin (I-T)(\partial\Omega)$. If $0 \notin (I-T)(\Omega)$, we define $deg(I-T, \Omega, 0) = 0$. Otherwise, put $F = \{x \in \overline{\Omega} : Tx = x\}$ and let C be the convex compact subset satisfying Proposition 3.1.17. Now, C is nonempty since $F \subset C$. Obviously, $T : C \cap \overline{\Omega} \to C$ is a mapping. If $r : E \to C$ is a retraction, then Tr is compact and $r^{-1}(\Omega)$ is open in E. By assumption, $0 \notin (I-T)(\partial\Omega)$ and we know that $0 \notin (I-Tr)(\partial(r^{-1}(\Omega) \cap \Omega)$, so the Leray Schauder degree $deg(I-Tr, r^{-1}(\Omega) \cap \Omega, 0)$ is well defined for each retraction r. Now, we define

$$deg(I-T, \Omega, 0) = deg(I - Tr, r^{-1}(\Omega) \cap \Omega, 0), \tag{3.2.1}$$

where $deg(I - Tr, r^{-1}(\Omega) \cap \Omega, 0)$ is the Leray Schauder degree.

To see this definition is reasonable, we show that, if $r_1, r_2 : X \to C$ are two retractions, then

$$deg(I - Tr_1, r_1^{-1}(\Omega) \cap \Omega, 0) = deg(I - Tr_2, r_2^{-1}(\Omega) \cap \Omega, 0).$$

Put $r(t,x) = tr_1 x + (1-t) r_2 x$ for all $(t,x) \in [0,1] \times E$. Then $r(t,\cdot) : E \to C$ is a retraction for each $t \in [0,1]$. Obviously, $x \neq Tr(t,x)$ for $(t,x) \in [0,1]\partial(r_1^{-1}(\Omega) \cap r_2^{-1}(\Omega) \cap \Omega)$. Thus, by the homotopy property of Leray Schauder degree, we have

$$\begin{aligned} &deg(I - Tr_1, r_1^{-1}(\Omega) \cap r_2^{-1}(\Omega) \cap \Omega, 0) \\ &= deg(I - Tr_2, r_1^{-1}(\Omega) \cap r_2^{-1}(\Omega) \cap \Omega, 0). \end{aligned}$$

It is simple to check that

$$0 \notin \overline{r_1^{-1}(\Omega) \cap \Omega} \setminus r_1^{-1}(\Omega) \cap r_2^{-1}(\Omega) \cap \Omega$$

and

$$0 \notin \overline{r_2^{-1}(\Omega) \cap \Omega} \setminus r_1^{-1}(\Omega) \cap r_2^{-1}(\Omega) \cap \Omega.$$

Thus the excision property of the Leray Schauder degree implies that

$$deg(I - Tr_1, r_1^{-1}(\Omega) \cap \Omega, 0) = deg(I - Tr_1, r_1^{-1}(\Omega) \cap r_2^{-1}(\Omega) \cap \Omega, 0)$$

and

$$deg(I - Tr_2, r_2^{-1}(\Omega) \cap \Omega, 0) = deg(I - Tr_2, r_1^{-1}(\Omega) \cap r_2^{-1}(\Omega) \cap \Omega, 0).$$

Therefore, we have

$$deg(I - Tr_1, r_1^{-1}(\Omega) \cap \Omega, 0) = deg(I - Tr_2, r_1^{-1}(\Omega) \cap r_2^{-1}(\Omega) \cap \Omega, 0).$$

One may also define a degree by taking the set C as in Proposition 3.1.18. This degree will coincide with the above one by the excision property of the Leray Schauder degree.

Theorem 3.2.1. The degree defined by 3.2.1 has the following properties:

(1) (*Normality*) $deg(I, \Omega, 0) = 1$ if and only if $0 \in \Omega$;

(2) (*Solvability*) If $deg(I - T, \Omega, 0) \neq 0$, then $Tx = x$ has a solution in Ω;

(3) (*Homotopy*) Let $H(t, x) : [0, 1] \times \overline{\Omega} \to E$ be a continuous and countably condensing mappings, i.e., $\alpha([0, 1] \times B) < \alpha(B)$ for all countable subset B of $\overline{\Omega}$ with $\alpha(B) > 0$ and $H(t, x) \neq x$ for all $(t, x) \in [0, 1] \times \partial\Omega$. Then $deg(I - H(t, \cdot), \Omega, 0)$ doesn't depend on $t \in [0, 1]$;

(4) (*Additivity*) Let Ω_1, Ω_2 be two disjoint open subsets of Ω and $0 \notin (I - T)(\overline{\Omega} - \Omega_1 \cup \Omega_2)$. Then

$$deg(I - T, \Omega, 0) = deg(I - T, \Omega_1, 0) + deg(I - T, \Omega_2).$$

Proof. (1), (2), and (4) follow directly from the definition and properties of the Leray Schauder degree.

To prove (3), we set

$$C_0 = \overline{conv(H([0, 1] \times \overline{\Omega}))}, \quad C_n = \overline{conv(H([0, 1] \times (C_{n-1} \cap \overline{\Omega})))}$$

for $n \geq 1$. Then $C = \cap_{n=0}^{\infty} C_n$ is compact by virtue of Proposition 3.1.18 and $H([0, 1] \times C) \to C$. Let $r : E \to C$ be a retraction. Then $x \neq H(t, rx)$ for all $x \in \partial(r^{-1}(\Omega \cap C) \cap \Omega)$. Thus $deg(I - H(t, \cdot)r, r^{-1}(C \cap \Omega) \cap \Omega, 0)$ doesn't depend on t. Therefore, the conclusion follows from the definition of the degree and the excision property of the Leray Schauder degree. This completes the proof.

Theorem 3.2.2. Let E be a Banach space, $\Omega \subset E$ be an open bounded subset with $0 \in \Omega$ and $T : \overline{\Omega} \to E$ be a continuous and countably condensing mapping. Suppose $x \neq \lambda Tx$ for all $\lambda \in [0,1), x \in \partial\Omega$. Then T has a fixed point in $\overline{\Omega}$.

Proof. We may assume that $Tx \neq x$ for all $x \in \partial\Omega$. Put $H(t,x) = tTx$ for all $(t,x) \in [0,1] \times \overline{\Omega}$. It is easy to see that $\{H(t,\cdot)\}_{t\in[0,1]}$ is a homotopy of countably condensing mappings. By assumption, we have $H(t,x) \neq x$ for all $x \in \partial\Omega$. As a result, $deg(I-T,\Omega,0) = deg(I,\Omega,0) = 1$. Thus $Tx = x$ has a solution in Ω, which is the desired result. This completes the proof.

Corollary 3.2.3. Let E be a Banach space, $\Omega \subset E$ be an open bounded subset with $0 \in \Omega$ and $T : \overline{\Omega} \to E$ be a continuous and countably condensing mapping. Suppose $\|Tx\| \leq \|x\|$ for all $x \in \partial\Omega$. Then T has a fixed point in $\overline{\Omega}$.

Proof. We may assume that $Tx \neq x$ for all $x \in \partial\Omega$. Otherwise, the conclusion is true. Thus we have $Tx \neq \lambda x$ for all $x \in \partial\Omega$ and $\lambda \leq 1$. By Theorem 3.2.2, T has a fixed point in $\overline{\Omega}$.

Theorem 3.2.4. Let E be a Banach space and $T : E \to E$ be a continuous and countably condensing mapping. Then one of the following conclusions holds:

(1) T has a fixed point in E;

(2) $\{x : Tx = \lambda x \text{ for some } \lambda > 1\}$ is unbounded.

Proof. Assume that $\{x : Tx = \lambda x \text{ for some } \lambda > 1\}$ is bounded. Take $r > 0$ such that $\{x : Tx = \lambda x \text{ for some } \lambda > 1\} \subset B(0,r)$. If $Tx = x$ for some $x \in \partial B(0,r)$, then (1) holds. So we may assume that $Tx \neq x$ for all $x \in \partial B(0,r)$. Thus $x \neq tTx$ for all $t \in [0,1]$ and $x \in \partial B(0,r)$, so $deg(I-T,B(0,r),0) = 1$ and T has a fixed point in $B(0,r)$. This completes the proof.

Theorem 3.2.5. Let E be a infinite dimensional Banach space, $\Omega \subset E$ be an open bounded subset with $0 \in \Omega$, $T : \overline{\Omega} \to E$ be a continuous and countably condensing mapping and $S : \partial\Omega \to E$ be a continuous compact mapping. Suppose $tTx + (1-t)Sx \neq x$ and $\|Sx\| \geq \|x\|$ for all $x \in \partial\Omega$ and $t \in [0,1]$. Then $deg(I-T,\Omega,0) = 0$.

Proof. First, there exists a continuous compact mapping $L : \overline{\Omega} \to E$ such that $Lx = Sx$ for all $x \in \partial\Omega$. Obviously, we have $tTx + (1-t)Lx \neq x$ for all $(t,x) \in [0,1] \times \partial\Omega$. Thus

$$deg(I-T,\Omega,0) = deg(I-L,\Omega,0).$$

By Lemma 2.2.11, we have $deg(I-L,\Omega,0) = 0$ and so $deg(I-T,\Omega,0) = 0$.

Corollary. 3.2.6. Let E be a infinite dimensional Banach space, $\Omega \subset E$ an open bounded subset with $0 \in \Omega$, $x_0 \in E$ such that $\|x_0\| > \sup_{x\in\partial\Omega} \|x\|$, and let $T : \overline{\Omega} \to E$ be a continuous and countably condensing mapping. Suppose $\|Tx\| > \|x\|$ and $Tx \neq \lambda x + (\lambda - 1)x_0$ for all $x \in \partial\Omega$, and $\lambda > 1$. Then $deg(I - T, \Omega, 0) = 0$. This completes the proof.

Proof. Define a mapping $S : \partial\Omega \to E$ by $Sx = x_0$. Then S is continuous, compact and $tTx + (1-t)Sx \neq x$ for all $x \in \partial\Omega$ and $t \in [0, 1]$. Obviously, $\|Sx\| > \|x\|$ for all $x \in \partial\Omega$. Thus the conclusion follows from Theorem 3.2.5.

Theorem 3.2.7. Let E be a infinite dimensional Banach space, $\Omega_i \subset E$, $i = 1, 2$, be two open bounded subsets such that $0 \in \Omega_1 \subset \Omega_2$, $x_0 \in E$ be such that $\|x_0\| > \sup_{x \in \partial\Omega_2} \|x\|$ and $T : \overline{\Omega_2} \to E$ be a continuous and countably condensing mapping. Suppose that the following conditions are satisfied:

(1) $\|Tx\| \leq \|x\|$ for all $x \in \partial\Omega_1$;

(2) $\|Tx\| \geq \|x\|$ and $Tx \neq \lambda x + (\lambda - 1)x_0$ for all $x \in \partial\Omega$ and $\lambda > 1$.

Then T has a fixed point in $\overline{\Omega_2} \setminus \Omega_1$.

Proof. We may assume that $Tx \neq x$ for $x \in \partial\Omega_1 \cup \partial\Omega_2$. By assumption (1), we know $x - tTx \neq 0$ for $t \in [0, 1]$ and $x \in \partial\Omega_1$. Thus $deg(I - T, \Omega_1, 0) = 1$.

On the other hand, by assumption (2) and Corollary 3.2.6, we have $deg(I - T, \Omega_2, 0) = 0$. Thus

$$deg(I - T, \Omega_2 \setminus \overline{\Omega_1}, 0) = deg(I - T, \Omega_2, 0) - deg(I - T, \Omega_1, 0) = -1$$

and so T has a fixed point in $\Omega_2 \setminus \overline{\Omega_1}$. This completes the proof.

Remark. One can define a degree theory for the so-called fundamentally restrictive mapping (see [291]).

3.3 Applications to ODEs in Banach Spaces

In this section, we give some applications to the ordinary differential equations in Banach spaces by using the degree theory for set contractive mappings.

Theorem 3.3.1. Let E be a Banach space and $f(t, x) : [0, 1] \times \overline{B(x_0, r_0)} \to E$ be a continuous mapping satisfying

$$\alpha(f([0, 1] \times B(x_0, r)) \leq k\alpha(B(x_0, r)) \quad \text{for all } r \in (0, r_0),$$

where $k \in (0, 1)$ and $r_0 > 0$ are constants. Then there exists $t_0 \in (0, 1]$ such that the initial value problem

$$\begin{cases} x'(t) = f(t, x(t)), & t \in (0, t_0), \\ x(0) = x_0 \end{cases} \qquad (E\ 3.3.1)$$

has a solution.

Proof. Set

$$M = \sup\{\|f(t,x)\| : (t,x) \in [0,1] \times \overline{B(x_0, r_0)}\},\ t_0 = \min\{1, \frac{r_0}{M}\}.$$

Obviously, $(E\ 3.3.1)$ is equivalent to the following integral equation:

$$x(t) = x_0 + \int_0^t f(s, x(s))ds. \qquad (E\ 3.3.2)$$

Put $X = C([0,t_0], E)$ with the norm $\|x(\cdot)\| = \max\{\|x(t)\| : t \in [0,t_0]\}$. Then X is a Banach space. We also set $K = \{x(\cdot) \in X : x(t_0) = x_0, \|x(t) - x(t_0)\| \leq r_0\}$. Then K is a bounded closed convex subset of X.

Now, we define a mapping $T : K \to K$ by

$$Tx(t) = x_0 + \int_0^t f(s, x(s))ds \quad \text{for all } x(\cdot) \in K.$$

It is easy to see that T is continuous. Now, we prove that T is condensing. In fact, for any subset B of K with $\alpha(B) > 0$, we have

$$\begin{aligned}
\alpha(TB) &= \sup_{t\in[0,t_0]} \alpha(\{Tx(t) : x(\cdot) \in B\}) \\
&= \sup_{t\in[0,t_0]} \alpha(\{x_0 + \int_0^t f(s, x(s)) : x(\cdot) \in B\}) \\
&\leq \sup_{t\in[0,t_0]} \alpha(\{x_0 + t\overline{conv(f(s,x(s)) : s \in [0,t_0])}\}) : x(\cdot) \in B\}) \\
&\leq \sup_{t\in[0,t_0]} \alpha(\{x_0 + t\overline{conv(f([0,t_0] \times B))}\}) \\
&\leq t_0 \sup_{t\in[0,t_0]} \alpha(f([0,t_0] \times B)) \\
&\leq t_0 k \alpha(B).
\end{aligned}$$

Thus T is a condensing mapping, so T has a fixed point in K, i.e., $(E\ 3.3.2)$ has a solution. Consequently, $(E\ 3.3.1)$ has a solution. This completes the proof.

Theorem 3.3.2. Let H be a real Hilbert space, $T > 0$ be a constant, $f(t,x) : R \times H \to H$ be a continuous mapping satisfying

$$\alpha(f([0,T] \times B) \leq k\alpha(B)$$

for all bounded subsets B of H, where $k \in (0,1)$ is a constant, and $kT < 1$. If $f(t+T, -x) = -f(t,x)$ and $\|f(t,x)\| \leq M\|x\| + g(t)$ for all $(t,x) \in R \times E$, where $0 \leq MT < 2$ is a constant, and $g(\cdot) \in L^2(0,T)$, then the following problem

$$\begin{cases} x'(t) = f(t, x(t)), & t \in R, \\ x(t+T) = -x(t) \end{cases} \qquad (E\ 3.3.3)$$

has a solution.

Proof. Let $C_a = \{x(\cdot) : R \to H$ is continuous $, x(t+T) = -x(t), t \in R\}$. Define $\|x(\cdot)\|_a = \max_{t\in[0,T]}$ for $x(\cdot) \in C_a$, and it is easy to check that C_a is a Banach space under this norm. It is simple to check that (E 3.3.3) is equivalent to the following equation:

$$x(t) = -\frac{1}{2}\int_t^T f(s,x(s))dt + \frac{1}{2}\int_0^t f(s,x(s))ds. \qquad (E\ 3.3.4)$$

We define a mapping $S : C_a \to C_a$ by

$$Sx(t) = -\frac{1}{2}\int_t^T f(s,x(s))dt + \frac{1}{2}\int_0^t f(s,x(s))ds \text{ for all } x(\cdot) \in C_a.$$

For any bounded subset U of C_a, we have, by Theorem 3.1.16, that

$$\alpha(SU) = \max_{t\in[0,T]}\{\alpha(\{Sx(t) : x(\cdot) \in U\})\}.$$

By the same reasoning as in Theorem 3.3.1, we get

$$\alpha(SU) \leq kT\alpha(U).$$

Thus S is a kT-set contraction.

Now, we prove that $Sx(\cdot) \neq \lambda x(\cdot)$ for all $\lambda > 1$ and $x(\cdot) \in C_a$ with

$$\|x(\cdot)\|_a > (1 - \frac{MT}{2})^{-1}\frac{\sqrt{T}}{2}(\int_0^T g^2(t)dt)^{\frac{1}{2}}.$$

In fact, suppose that there exists $x(\cdot) \in C_a$ with

$$\|x(\cdot)\| > (1 - \frac{MT}{2})^{-1}\frac{\sqrt{T}}{2}(\int_0^T g^2(t)dt)^{\frac{1}{2}}$$

such that $Sx(\cdot) = \lambda x(\cdot)$. Then we have

$$\lambda x'(t) = f(t,x(t)). \qquad (3.3.1)$$

Multiply both sides of (3.3.1) by $x'(t)$ and integrate over $[0,T]$, we have

$$\lambda\int_0^T \|x'(t)\|^2 dt = \int_0^T f(t,x(t))x'(t)dt$$

$$\leq M\int_0^T \|x(t)\|\|x'(t)\|dt + \int_0^T g(t)\|x'(t)\|dt.$$

In view of Lemma 2.5.4, we have

$$\lambda\int_0^T \|x'(t)\|^2 dt \leq \frac{MT}{2}\int_0^T \|x'(t)\|^2 dt + (\int_0^T g^2(t)dt)^{\frac{1}{2}}(\int_0^T \|x'(t)\|^2 dt)^{\frac{1}{2}},$$

i.e.,

$$(\int_0^T \|x'(t)\|^2 dt)^{\frac{1}{2}} \leq (1 - \frac{MT}{2})^{-1} (\int_0^T g^2(t) dt)^{\frac{1}{2}}.$$

Again, by Lemma 2.5.4, we get

$$\|x(\cdot)\|_a \leq (1 - \frac{MT}{2})^{-1} \frac{\sqrt{T}}{2} (\int_0^T g^2(t) dt)^{\frac{1}{2}},$$

which is a contradiction. Thus by Theorem 3.2.4, S has a fixed point in C_a, i.e., the problem (E 3.3.3) has a solution. This completes the proof.

3.4 Exercises

1. Let $A_i \subset C([0,1])$, $i = 1, 2, 3$, be defined by

   $A_1 = \{x(\cdot) : x(0) = 0,\ x(1) = 1 \text{ and } 0 \leq x(t) \leq 1 \text{ for } t \in [0,1]\}$;

   $A_2 = \{x(\cdot) : 0 \leq x(t) \leq \frac{1}{2},\ t \in [0, \frac{1}{2}],\ \frac{1}{2} \leq x(t) \leq 1,\ t \in [\frac{1}{2}, 1]\} \cap A_1$;

   $A_3 = x(\cdot) : 0 \leq x(t) \leq \frac{2}{3},\ t \in [0, \frac{1}{2}],\ \frac{1}{3} \leq x(t) \leq 1,\ t \in [\frac{1}{2}, 1]\} \cap A_1$;

   Show $\beta(A_1) = \frac{1}{2}$, $i = 1, 2, 3$, and $\alpha(A_1) = 1$, $\alpha(A_2) = \frac{1}{2}$, $\alpha(A_3) = \frac{2}{3}$.

2. Let E be a Banach space, $C^1([a,b], E)$ be the space of continuously differentiable functions with the norm

   $$\|x(\cdot)\|_1 = \max_{t \in [a,b]} \|x(t)\| + \max_{t \in [a,b]} \|x'(t)\|$$

   and α_1 is the Kuratowski measure of noncompactness in $C^1([a,b], E)$. Let $B \subset C^1([a,b], E)$ be a bounded subset such that $B' = \{x'(\cdot) : x(\cdot) \in B\}$ is equicontinuous. Show that

   $$\alpha_1(B) = \max\{\max_{t \in [a,b]} \alpha(\{x(t) : x(\cdot) \in B\}),\ \max_{t \in [a,b]} \alpha(\{x'(t) : x(\cdot) \in B\})\}.$$

3. Let X be a complete metric space and $A_i \subset X$ be a closed subset for $i = 1, 2, \cdots$. Suppose that $A_1 \supset A_2 \supset A_3 \supset \cdots$, and $\lim_{n \to \infty} \phi(A_n) = 0$, where ϕ is the Kuratowski measure or the Hausdorff measure of noncompactness. Show that $\cap_{n=1}^{\infty} A_n \neq \emptyset$.

4. Let X be a Banach space and $T : X \to X$ be a continuous and compact mapping. Suppose that there exists a linear bounded mapping $L : X \to X$ such that $\lim_{\|x\| \to \infty} \frac{\|Tx - Lx\|}{\|x\|} = 0$, i.e., T is asymptotically linear. Show that L is compact.

5. Let B be the closed unit ball of l^2 and $T : B \to B$ is defined by

$$T(x_1, x_2, \cdots) = (\sqrt{1 - \|x\|^2}, x_1, x_2, \cdots)$$

for all $x = (x_1, x_2, \cdots) \in B$. Show that T is an 1-set contraction.

6. Let X be a real Banach space, $B(0, R) \subset X$ be the open ball of origin with radius R and $T : \overline{B(0, R)} \to X$ be a condensing mapping. Suppose that one of the following condition is satisfied:

   (1) $T(\partial B(0, R)) \subset \overline{B(0, R)}$;

   (2) $\|Tx - x\|^2 \geq \|Tx\|^2 - \|x\|^2$ for all $x \in \partial B(0, R)$.

   Show that T has a fixed point in $\overline{B(0, R)}$.

7. Let X be a real Banach space, $\Omega \subset X$ be an open bounded subset with $0 \in \Omega$ and $T : \overline{\Omega} \to X$ be an 1-set contraction such that $(I - T)(\overline{\Omega})$ is closed and, if $Tx = \lambda x$ for all $x \in \partial\Omega$, then $\lambda \leq 1$. Show that T has a fixed point in $\overline{\Omega}$.

8. Let X be a Banach space with a Schauder basis $\{e_i : i \in N\}$ for each $x = \Sigma_{i=1}^{n} a_i(x)e_i$ and $R_n x = \Sigma_{i=n+1}^{\infty} a_i(x)e_i$ for $n = 1, 2, \cdots$. For each bounded subset $B \subset X$, define $\mu(B) = \limsup_{n\to\infty} \sup\{\|R_n x\| : x \in B\}$. Show that μ has the following properties:

   (1) $\mu(B) = 0$ if and only if B is relatively compact;

   (2) $\mu(A \cup B) = \max\{\mu(A), \mu(B)\}$;

   (3) $\mu(conv(A)) = \mu(A)$.

9. Let X be a real Banach space, $C \subset X$ be a bounded closed convex subset and $T : C \to C$ be an 1-set contraction. Show that $\inf_{x\in C}\{\|x - Tx\|\} = 0$.

10. Let X be a real Banach space, $\Omega \subset X$ an open bounded subset and $T : \Omega \to X$ be a k-set contraction with $k \in (0, 1)$. Suppose that $I - T$ is a homeomorphism from $\Omega \to (I - T)(\Omega)$. Show that $\alpha([I - (I - T)^{-1}]B) \leq k(1 - k)^{-1}\alpha(B)$ for all bounded subset B of Ω.

11. Let X be a real Banach space, $\Omega \subset X$ be an open bounded subset with $0 \in \Omega$, $T : \overline{\Omega} \to X$ be a countably condensing mapping and $A : D(A) \subseteq E \to 2^E$ be a $m-$accretive mapping such that $0 \in A0$. Suppose that $\|Tx\| \leq \|x\|$ for all $x \in \partial\Omega$. Show that $-A + T$ has a fixed point in $\overline{\Omega} \cap D(A)$.

12. Let X be a complete metric space, and $A_i \subset X$ a clsoed subset for $i = 1, 2, \cdots$. Suppose that $A_1 \supset A_2 \supset A_3 \supset \cdots$, and $\lim_{n\to\infty} \alpha(A_n) = 0$, $A_\infty = \cap_{n=1}^{\infty} A_n$. Show $\lim_{n\to\infty} H(A_n, A_\infty) = 0$.

13. Show that the equation

$$\begin{cases} \frac{\partial u(s,t)}{\partial s} = \frac{1}{2}u(s,2t) + \frac{1}{2}u(s,0) + \sin \pi s, t \in [0,\frac{1}{2}], & s \in R \\ \frac{\partial u(s,t)}{\partial s} = \frac{1}{2}u(s,2t-1) + \frac{1}{2}u(s,1) + \sin \pi s, & t \in [\frac{1}{2},1],\ s \in R \\ u(s+1,t) = -u(s,t), & t \in [0,1],\ s \in R, \end{cases}$$

has a solution $u(s,t)$ such that $u(s,\cdot) \in C([0,1])$ for all $s \in R$.

Chapter 4

GENERALIZED DEGREE THEORY FOR A-PROPER MAPS

To solve an infinite dimensional equation $Tx = y$, a very natural method is to approximate the original equation by finite dimensional equations, as we have seen in the Leray Schauder theory. The well-known Galerkin method has proved to be a very efficient tool in finite dimensional approximation. In the 1960s, Browder and Petryshyn systematically studied the finite dimensional method for a large class of mappings, which they called A-proper mappings, and they developed a similar theory to the Brouwer degree.

Our goal of this chapter is to introduce Petryshyn's generalized degree theory for A-proper mappings. This chapter has five main sections.

In Section 4.1, we define projection schemes and A-Proper mappings. Various examples (see lemmas 4.1.8 and 4.1.10) are also discussed.

Section 4.2 presents a degree theory for A-proper mappings and various properties are presented (see Theorem 4.2.3 and Proposition 4.2.5).

A variety of existence results are presented in Section 4.3 for the semilinear situation $S - L$ where L is a Fredholm map of index zero and $S - \lambda N$ is A-Proper for each $\lambda \in (0, 1]$, where N is a mapping satisfying some specific conditions (see theorems 4.3.3 and 4.3.4).

In Section 4.4, we introduce the notion of a Fredholm mapping of index zero type. We present a degree (coincidence) theory for maps $L - N$ where L is a Fredholm mapping of index zero type and N is such that either (i) N is L-A-Proper or (ii) $L + \lambda JP - N$ is A-proper for all $\lambda \in (0, \lambda_0)$ with $\lambda_0 > 0$ or (iii) $I - (L + \lambda JP)^{-1}(N + \lambda JP)$ is A-proper for some $\lambda > 0$ hold.

The results of Section 4.3 and 4.4 are used in Section 4.5 to present the existence results for the periodic semilinear ordinary and partial differential equations.

4.1 A-Proper Mappings

Definition 4.1.1. Let X and Y be real separable Banach spaces.

(1) If there is a sequence of finite dimensional subspaces $X_n \subset X$ and a

sequence $\{P_n\}$ of linear projections $P_n : X \to X_n$ such that $P_n x \to x$ for all $x \in X$, then we say that X has a projection scheme $\{X_n, P_n\}$.

(2) If X and Y have projection schemes $\{X_n, P_n\}$ and $\{Y_n, Q_n\}$, respectively, and $dim X_n = dim Y_n$ for all positive integers n, then we call $\Pi = \{X_n, P_n; Y_n, Q_n\}$ an operator projection scheme.

Example 4.1.2. Let $X = C([0,1])$. For $n \in N$, partition $[0,1]$ into n equal parts and set $t_0 = 0 < t_1 = \frac{1}{n} < t_2 = \frac{2}{n} < \cdots < t_n = 1$. Let X_n be the subspace of all $x \in X$ which are linear in every subinterval $[t_i, t_{i+1}]$ and $P : X \to X_n$ be the projection satisfying $P_n x(t_i) = x(t_i)$ for $i = 1, 2, \cdots, n$. Then $\{X_n, P_n\}$ is a projection scheme for X.

Example 4.1.3. Let X be a Banach space with a Schauder basis $\{e_i : i \in N\}$. Then X has a projection scheme $\{X_n, P_n\}$ defined by

$$X_n = span\{e_1, e_2, \cdots, e_n\}, \quad P_n x = \Sigma_{i=1}^n \alpha_i(x) e_i$$

for $x = \Sigma_{i=1}^\infty \alpha_i(x) e_i$. In the case of a separable Hilbert space, we may choose an orthonormal basis $\{e_i : i \in N\}$, then the projection $P_n x = \Sigma_{i=1}^n (x, e_i) e_i$ satisfies $P_n^* = P_n$ and $\|P_n\| = 1$.

Example 4.1.4. Let X be a reflexive Banach space with a projection scheme such that $P_n P_m = P_{\min\{m,n\}}$. Then $\{P_n^* X^*, P_n^*\}$ is a projection scheme for X^*.

Proof. Notice that $P_n^* P_n^* f(x) = f(P_n^2 x) = P_n^* f(x)$ on X and thus P_n^* is a projection. We also have

$$dim P_n^* X = dim N(I^* - P^*) = dim N(I - P_n) = dim X_n.$$

We claim that $X^* = \overline{\cup_{i=1}^\infty P_i^* X^*}$. If not, there is $x_0 \in X \setminus \{0\}$ such that $f(x) = 0$ for all $f \in \overline{\cup_{i=1}^\infty P_i^* X^*}$ since $X^{**} = J(X)$, where $J(x)(f) = f(x)$ for all $f \in X^*$ and $x \in X$. Thus, we have $f(P_n x) = 0$ for all n and $f \in X^*$, so $f(x) = 0$ for all $f \in X^*$ and so $x = 0$, which is a contradiction. Therefore, $X^* = \overline{\cup_{i=1}^\infty P_i^* X^*}$. We also have $P_n^* X^* \subset P_m^* X^*$ for $n < m$. Thus, for any $f \in X^*$ and $\epsilon > 0$, we may choose $g \in P_n^* X^*$ such that $\|f - g\| < \epsilon$ and we then have

$$\|P_m f - f\| \leq \|P_m(f - g)\| + \|g - f\| \leq (\sup_{n \geq 1} \|P_n\| + 1)\epsilon,$$

which gives $P_m^* f \to f$ as $m \to \infty$.

Example 4.1.5. If both X and Y have Schauder basis, then there exists an operator projection scheme.

Proof. Let $\{e_n\}$ be a Schauder basis of X and $\{e_n'\}$ be a Schauder basis of Y. Put $X_n = span\{e_1, e_2, \cdots, e_n\}$ and $Y_n = span\{e_1', e_2', \cdots, e_n'\}$. For

$x = \Sigma_{i=1}^{\infty}\alpha_i e_i$ and $y = \Sigma_{i=1}^{\infty}\beta_i e_i'$, set $P_n x = \Sigma_{i=1}^{n}\alpha_i e_i$ and $Q_n y = \Sigma_{i=1}^{n}\beta_i e_i'$. Then $\Pi = \{X_n, P_n; Y_n, Q_n\}$ is an operator projection scheme.

Definition 4.1.6. Let X, Y be real Banach spaces and $\Pi = \{X_n, P_n; Y_n, Q_n\}$ be an operator projection scheme. Then a mapping $T : D \subset X \to Y$ is called A- proper (respectively, pseudo A-proper) with respect to Π if, for any bounded $x_m \in D \cap X_m$ and $Q_m T x_m \to y$, there exists a subsequence $\{x_{m_k}\}$ such that $x_{m_k} \to x \in D$ and $Tx = y$, (respectively, there exists $x \in D(T)$, such that $Tx = y$). We denote by $A_\Pi(D, Y)$ the class of all A-proper mappings $F : D \to Y$.

Recall that, in a normed space X, the semi-inner products $(\cdot,\cdot)_-$ and $(\cdot,\cdot)_+$ are defined by

$$(x, y)_- = \lim_{h\to 0^+} h^{-1}(\|x\| - \|x - hy\|) \quad \text{for all } x, y \in X,$$

$$(x, y)_+ = \lim_{h\to 0^+} h^{-1}(\|x + hy\| - \|x\|) \quad \text{for all } x, y \in X.$$

An operator $T : D(T) \subseteq X \to X$ is called *accretive* if

$$(x - y, Tx - Ty)_+ \geq 0 \quad \text{for all } x, y \in D(T)$$

and *strongly accretive* if

$$(x - y, Tx - Ty)_+ \geq c\|x - y\|^2 \quad \text{for some constant } c > 0.$$

For some properties of accretive operators, we refer the reader to [17] and [60]. One can show that a continuous strongly accretive operator $T : X \to X$ is A-proper, and the proof is left to the reader as an exercise.

Definition 4.1.7. Let X be a separable Banach space with a projection scheme $\Pi = \{X_n, P_n\}$. Then $T : D \subseteq X \to X$ is called a P_1 compact mapping if $\lambda I - T$ is A-proper with respect to Π for all $\lambda \geq 1$.

In the following, let X, Y be separable Banach spaces, $S : X \to Y$ be a linear Fredholm mapping of index zero with $N(S) \neq \{0\}$ and $C : D \subset X \to Y$ be a nonlinear mapping. Consider the semilinear problem $Sx - Cx = y$ for all $x \in D(L) \cap D$ and $y \in Y$. Since S is Fredholm of index zero (see Chapter V), there exist closed subspaces X' of X and Y' of Y with $dimY' = dimN(S)$ such that

$$X = N(S) \oplus X', \quad Y = Y' \oplus R(S).$$

Let $P : X \to N(S)$ be a projection, $Q : Y \to Y'$ be a projection and $M : N(S) \to Y'$ be a isomorphism. Put $T = MP$. Then T is a compact linear operator. It is known that $S + T$ is also a Fredholm mapping with $ind(S + T) = ind(S) = 0$ and $S + T$ is bijective with $(S + T)^{-1} : Y \to X$ bounded. Set $S_1 = S|_{X'\cap D(S)}$. Then S_1 is injective and closed and so S_1^{-1} is continuous on $R(S)$.

Suppose Y has a sequence of finite dimensional subspaces Y_n with a sequence of projections $Q_n : Y \to Y_n$ such that $Q_n y \to y$ for all $y \in Y$ as $n \to \infty$. If $S : X \to Y$ is a Fredholm mapping of index zero, we set $X_n = (S+T)^{-1}Y_n$ and then $\Gamma_S = \{X_n, Y_n, Q_n\}$ is an admissible scheme for (X, Y).

Lemma 4.1.8. Let $S : X \to Y$ be a Fredholm mapping of index zero. Then S is A-proper with respect to Γ_S.

Proof. Obviously, $Q_n S : X_n \to Y_n$ is continuous. Let $(x_{n_j}) \in X_{n_j}$ be a bounded sequence such that $Q_{n_j} S x_{n_j} \to y \in Y$. Since $Q_n(S+T)x = (S+T)x$ for all $x \in X_n$, we have

$$Q_{n_j}(S+T)x_{n_j} = (S+T)x_{n_j}.$$

Since T is compact, we may assume that $Tx_n \to z$ by taking a subsequence. Therefore, we have

$$x_{n_j} = (S+T)^{-1}[Q_{n_j} S x_{n_j} + Q_{n_j} T x_{n_j}] \to x = (S+T)^{-1}(y+z)$$

and $Tx = z$. Thus $Sx = y$, i.e., S is A-proper with respect to Π_S.

Definition 4.1.9. Let X, Y be real Banach spaces, $\Pi = \{X_n, P_n; Y_n, Q_n\}$ be an operator projection scheme and $D \subset X$. A family of mappings $H(t,x) : [0,1] \times D \to Y$ is called A-proper homotopy with respect to Π if

(1) for any bounded sequence (x_m) in $D \cap X_m$, $t_m \to t_0$ and $Q_m H(t_m, x_m) \to y$, there exists a subsequence (x_{m_k}) of (x_m) such that $x_{m_k} \to x \in D$ and $H(t_0, x) = y$;

(2) $Q_n H(t,x) : [0,1] \times D \cap X_n \to Y_n$ is continuous for $n = 1, 2, \cdots$.

If $S : D(S) \subset X \to Y$ is a Fredholm mapping, then it is known from [283] that

$$l(S) = sup\{r \in R^+ : r\gamma(B) \le \gamma(S(B)),\ B \subset D(S) \text{ is bounded}\} > 0.$$

Lemma 4.1.10. Suppose that S is Fredholm of index zero, $\Omega \subset X$ is an open bounded subset, Γ_S is as above, $N : \overline{\Omega} \cap D(S) \to Y$ is a bounded continuous mapping and $T_\lambda = S - \lambda N$ for $\lambda \in (0,1]$. Assume that one of the following conditions holds:

(1) N or $S_1^{-1} : R(S) \to X$ is compact;

(2) N is k-ball contractive with $k \in [0, l(S))$ and $\|Q_n\| = 1$;

(3) $N(S+T)^{-1} : (S+T)(\overline{\Omega} \cap D(S)) \to Y$ is ball condensing and $\|Q_n\| = 1$.

Then T_λ is A-proper with respect to Γ_S for each $\lambda \in (0, 1]$.

Proof. (1) If N is compact, then it is trivial that $S_\lambda N$ is A-proper with respect to Π_S. Suppose now that S_1^{-1} is compact. Let $(x_{n_j}) \in \overline{\Omega} \cap X_{n_j}$ be any sequence such that

$$g_{n_j} = Q_{n_j} S x_{n_j} - \lambda Q_{n_j} N x_{n_j} \to g \in Y.$$

Since $Q_n(S+T)x = (S+T)x$ for all $x \in X_n$, we have

$$g_{n_j} = (S+T)x_{n_j} - \lambda Q_{n_j} N x_{n_j} - Q_{n_j} T x_{n_j} \to g.$$

But T is compact, so we have $Sx_{n_j} - \lambda Q_{n_j} N x_{n_j} \to g$. Therefore, it follows that

$$(I-Q)Sx_{n_j} - \lambda(I-Q)Q_{n_j} N x_{n_j} \to (I-Q)g \in R(S).$$

Thus we have

$$(I-P)x_{n_j} - \lambda S_1^{-1}(I-Q)Q_{n_j}(Nx_{n_j}) \to S_1^{-1}(I-Q)g.$$

Since P and $S_1^{-1}(I-Q)$ are compact, it is easy to see that $(x_{n_j})_{j=1}^{\infty}$ has a subsequence $(x'_{n_j})_{j=1}^{\infty}$ converging to x_0. From the continuity of N, it follows that $Sx'_{n_j} \to g + \lambda N x_0$. The closedness of S implies that

$$Sx_0 - \lambda N x_0 = g$$

and thus $S - \lambda N$ is A-proper.

(2) Let $x_{n_j} \in \overline{\Omega} \cap X_{n_j}$ be any bounded sequence such that $Q_{n_j} S x_{n_j} - \lambda Q_{n_j} N x_{n_j} \to g$. As in (1), we have $Sx_{n_j} - \lambda Q_{n_j} N x_{n_j} \to g$.

On the other hand, $\gamma(\{Sx_{n_j}\}) \leq \lambda k \gamma(\{x_{n_j}\}) \leq k\gamma(\{x_{n_j}\})$. This and the assumption that $k < l(S)$ imply that $\gamma(\{x_{n_j}\}) = 0$, so we may assume that $x_{n_j} \to x_0$ and, consequently, $Sx_{n_j} \to g + \lambda N x_0$. Hence $Sx_0 - \lambda N x_0 = g$ and so $S - \lambda N$ is A-proper.

(3) Let (x_{n_j}) be any bounded sequence in $\overline{\Omega} \cap X_{n_j}$ such that $Q_{n_j} S x_{n_j} - \lambda Q_{n_j} N x_{n_j} \to g$. Then we have

$$(S+T)x_{n_j} - \lambda Q_{n_j} N x_{n_j} - Q_{n_j} T x_{n_j} \to g.$$

Set $y_{n_j} = (S+T)x_{n_j}$. We have

$$y_{n_j} - \lambda Q_{n_j} N(S+T)^{-1}(y_{n_j}) - Q_{n_j} T(S+T)^{-1}(y_{n_j}) \to g.$$

By the compactness of $T(S+T)^{-1}$ and the assumption (3), it follows that $(y_{n_j})_{j=1}^{\infty}$ has a subsequence (y'_{n_j}) with $y'_{n_j} \to y_0$. Thus $x'_{n_j} = (S+T)^{-1} y_{n_j} \to x_0$, and it is simple to check that $Sx_0 - \lambda N x_0 = g$. This completes the proof.

4.2 Generalized Degree for A-Proper Mappings

Let X, Y be real separable Banach spaces and $\Pi = \{X_n, P_n; Y_n, Q_n\}$ be an operator projection scheme. Let $\Omega \subset X$ be an open bounded subset and L be a dense subspace of X with $\cup_{n=1}^{\infty} X_n \subset L$. We let $\Omega_L = \Omega \cap L$.

Lemma 4.2.1. Let $T \in A_\Pi(\overline{\Omega} \cap L, Y)$. Suppose that $p \notin T(\partial\Omega \cap L)$. Then there exists an integer $n_0 > 0$ such that

$$Q_n p \notin Q_n T(\partial(\Omega \cap X_n)) \text{ for all } n > n_0.$$

Proof. Suppose that the assertion of Lemma 4.2.1 is not true. Then there exists $n_k \to \infty$ and $x_{n_k} \in \partial\Omega \cap X_{n_k}$ such that $Q_{n_k} p = Q_{n_k} T x_{n_k}$. Obviously, $x_{n_k} \in \partial\Omega \cap L$. Thus we have $Q_{n_k} T x_{n_k} \to p$ as $k \to \infty$ and the A-properness of T guarantees the existence of a subsequence $(x_{n_{k_l}})_{l=1}^{\infty}$ such that $x_{n_{k_l}} \to x_0 \in \partial\Omega \cap L$, and $Tx_0 = p$, which is a contradiction. This completes the proof.

Definition 4.2.2. Let $T \in A_\Pi(\overline{\Omega} \cap L, Y)$. Suppose that $p \notin T(\partial\Omega \cap L)$ and $Q_n T$ is continuous. We define a generalized degree $D(T, \Omega, p)$ by

$$Deg(T, \Omega \cap L, p) = \{k \in Z \cup \{\pm\infty\} : deg(Q_{n_j} T, \Omega \cap X_{n_j}, Q_{n_j} p) \to k \text{ for some } n_j \to \infty\},$$

where Z is the set of all integers.

By Lemma 4.2.1, we know that there exists an integer $n_0 > 0$ such that $p \notin Q_n T(\partial\Omega \cap X_n)$ and $Q_n T$ is continuous, so the Brouwer degree $deg(Q_n T, \Omega \cap X_n, Q_n p$ is well defined for $n > n_0$. Thus $Deg(T, \Omega \cap L, p)$ is nonempty and the definition is well defined.

Theorem 4.2.3. Let $T \in A_\Pi(\overline{\Omega} \cap L, Y)$ and $p \notin T(\partial\Omega \cap L)$. Then the generalized degree has the following properties:

(1) If $Deg(T, \Omega \cap L, p) \neq \{0\}$, then $Tx = p$ has a solution in $\Omega \cap L$;

(2) If $\Omega_i \subset \Omega$ for $i = 1, 2$, $\Omega_1 \cap \Omega_2 = \emptyset$ and $p \notin (\overline{\Omega} \setminus \Omega_1 \cup \Omega_2) \cap L$, then

$$Deg(T, \Omega \cap L, p) \subseteq Deg(T, \Omega_1 \cap L, p) + Deg(T, \Omega_2 \cap L, p),$$

here we use the convention that $+\infty + (-\infty) = Z \cup \{\pm\infty\}$;

(3) If $H(t, x) : [0, 1] \times \overline{\Omega} \cap L \to Y$ is a A-proper homotopy and $p \notin H(t, x)$ for all $(t, x) \in [0, 1] \times \partial\Omega \cap L$, then $Deg(H(t, \cdot), \Omega \cap L, p)$ does not depend on $t \in [0, 1]$;

(4) If $0 \in \Omega$, Ω is symmetric about 0, $T : \overline{\Omega} \cap L \to Y$ is an odd A-proper mapping and $0 \notin T(\partial\Omega \cap L)$, then $Deg(T, \Omega \cap L, 0)$ contains no even numbers.

Proof. (1) If $Deg(T, \Omega \cap L, p) \neq \{0\}$, then there exists $n_k \to \infty$ such that

$$deg(Q_{n_k} T, \Omega \cap X_{n_k}, Q_{n_k} p) \neq 0.$$

Thus there exists $x_{n_k} \in \Omega \cap L$ such that $Q_{n_k} T x_{n_k} = Q_{n_k} p$. By the A-properness of T, there is a subsequence $(x_{n_{k_j}})$ with $x_{n_{k_j}} \to x_0 \in \Omega \cap L$ and $Tx_0 = p$.

(2) Since $p \notin (\overline{\Omega} \setminus \Omega_1 \cup \Omega_2) \cap L$, there exists $n_0 > 0$ such that

$$Q_n p \notin (\overline{\Omega} \setminus \Omega_1 \cup \Omega_2) \cap X_n \quad \text{for all } n > n_0.$$

Therefore, we have

$$deg(Q_n T, \Omega, p) = deg(Q_n T, \Omega_1, p) + deg(Q_n T, \Omega_2, p) \text{ for all } n > n_0.$$

If $k = \lim_{j\to\infty} deg(Q_{n_j} T, \Omega \cap X_{n_j}, Q_{n_j} p)$, then we have

$$k = \lim_{j\to\infty} [deg(Q_{n_j} T, \Omega_1 \cap X_{n_j}, Q_{n_j} p) + deg(Q_{n_j} T, \Omega_2 \cap X_{n_j}, Q_{n_j} p).$$

If $\lim_{j\to\infty} deg(Q_{n_j} T, \Omega_1 \cap X_{n_j}, Q_{n_j} p)$ and $\lim_{j\to\infty}[deg(Q_{n_j} T, \Omega_2 \cap X_{n_j}, Q_{n_j} p)$ are both equal to $+\infty$ or $-\infty$, then $k = +\infty$ or $k = -\infty$ and so the conclusion holds. If one of them equals to $+\infty$ and the other one is $-\infty$, then, by the convention, we have $k \in Z \cup \{\pm\infty\}$. For the case,

$$\limsup_{j\to\infty} |deg(Q_{n_j} T, \Omega_1 \cap X_{n_j}, Q_{n_j} p)| < +\infty$$

and

$$\limsup_{j\to\infty} |deg(Q_{n_j} T, \Omega_2 \cap X_{n_j}, Q_{n_j} p)| < +\infty$$

and so the conclusion is obvious.

(3) We claim that there exists an integer $n_0 > 0$ such that

$$Q_n p \notin \cup_{t\in[0,1]} H(t, \partial\Omega \cap X_n) \quad \text{for all } n > n_0.$$

Assume the assertion is false. Then there exist $n_j \to \infty$, $t_j \to t_0$, $x_{n_j} \in \partial\Omega \cap X_{n_j}$ such that $Q_{n_j} p = Q_{n_j} H(t_j, x_{n_j})$. Therefore, (x_{n_j}) has a subsequence (x'_{n_j}) which converges to $x_0 \in \partial\Omega \cap L$ and $H(t_0, x_0) = p$, which is a contradiction. Thus the Brouwer degree $deg(Q_n H(t, \cdot), \Omega \cap X_n, Q_n p)$ does not depend on $t \in [0, 1]$ for $n > n_0$, so $Deg(H(t, \cdot), \Omega \cap L, p)$ does not depend on $t \in [0, 1]$.

(4) Since $\Omega \cap X_n$ is symmetric about 0, by Borsuk's theorem, we have $deg(Q_n T, \Omega \cap X_n, 0)$ is odd for n sufficiently large. Thus $Deg(T, \Omega \cap L, 0)$ contains no even numbers. This completes the proof.

Corollary 4.2.4. Let X, Y be separable Banach spaces, $S : D(S) \subseteq X \to Y$ be a Fredholm mapping of index zero and (Y_n, Q_n) be a projection scheme for Y. If Ω is an open bounded subset of X with $\overline{\Omega} \cap D(L) \neq \emptyset$ and $N : \overline{\Omega} \cap D(S) \to Y$ is a nonlinear mapping such that $S - N$ is A-proper with respect to Γ_S, then we have the following:

(1) If $p \notin (S - N)(\partial\Omega \cap D(S))$ and $Deg(S - N, \Omega \cap L, p) \neq \{0\}$, then $Sx - Nx = p$ has a solution in $\Omega \cap D(S)$;

(2) If Ω is symmetric about 0, N is odd and $0 \notin (S - N)(\partial\Omega \cap L)$, then $Deg(S - N, \Omega \cap D(S), 0)$ contains no even numbers.

Proof. The proof follows from Theorem 4.2.3.

Proposition 4.2.5. Let X, Y be separable Banach spaces, $S : D(S) \subseteq X \to Y$ be a Fredholm mapping of index zero and (Y_n, Q_n) be a projection scheme for Y. If Ω is an open bounded subset of X with $\overline{\Omega} \cap D(L) \neq \emptyset$ and $A : X \to Y$ is a linear continuous compact mapping, then $S - A$ is A-proper with respect to Γ_S. If $N(S - A) = \{0\}$, then

$$Deg(S - A, \Omega \cap D(S), 0) = \begin{cases} \{0\}, & 0 \notin \overline{\Omega}, \\ \{1\} \text{ or } \{-1\}, & 0 \in \Omega. \end{cases}$$

Proof. It is easy to see that $S - A$ is A-proper with respect to Π_S. Since $N(S - A) = \{0\}$, $S - A$ is injective, it follows that $Sx - Ax \neq 0$ for all $x \in \partial\Omega \cap D(S)$. Thus, if $0 \notin \overline{\Omega}$, then $Deg(S - A, \Omega \cap D(S), 0) \neq 0$ would imply that $Sx - Ax = 0$ has a solution in $\Omega \cap D(S)$, which contradicts $N(S - A) = \{0\}$.

On the other hand, if $0 \in \Omega$, then $0 \in \Omega \cap X_n$ for all n, $Q_n(S - A) : X_n \to Y_n$ is injective for sufficiently large n and we have

$$deg(Q_n(S - A), \Omega \cap X_n, 0) = 1 \text{ or } -1.$$

This completes the proof.

4.3 Equations with Fredholm Mappings of Index Zero

In this section, all the notations are the same as in previous sections.

Proposition 4.3.1. Let X, Y be separable Banach spaces, $S : D(S) \subseteq X \to Y$ be a Fredholm mapping of index zero and (Y_n, Q_n) be a projection scheme for Y. Suppose that $\Omega \subset X$ is symmetric about 0. Let $N : \overline{\Omega} \cap D(S) \to Y$ be a mapping such that

(1) $H(t,x) = Sx - [\frac{1}{1+t}N(x) - \frac{t}{1+t}N(-x)]$ is A-proper with respect to Γ_S for all $t \in [0,1]$;

(2) $\|N(x) + N(-x)\| \le d$ for all $x \in \partial\Omega \cap D(S)$ and some $d > 0$;

(3) $S(x) - N(x) \ne \lambda[S(-x) - T(-x)]$ for all $x \in \partial\Omega \cap D(S)$ and $\lambda \in [0,1]$.

Then $Sx - Nx = 0$ has a solution in $D(S)$.

Proof. By (1) and (2), it is simple to check that $(H(t,\cdot))_{t\in[0,1]}$ is an A-proper homotopy. Moreover, in view of (3), $H(t,x) \ne 0$ for all $(t,x) \in [0,1]\times\partial\Omega\cap D(S)$. Thus $Deg(H(t,\cdot),\Omega\cap D(S),0)$ does not depend on $t \in [0,1]$. But $H(1,x) = Sx - \frac{1}{2}[N(x) - N(-x)]$ is an odd mapping, so $Deg(H(1,\cdot),\Omega\cap D(S),0)$ contains no even numbers. Thus $Deg(L-N,\Omega\cap D(S),0)$ contains no even numbers, so $Lx - Nx$ has a solution in $\Omega \cap D(S)$. This completes the proof.

Proposition 4.3.2. Assume that the following conditions hold:

(a) $S - \lambda N : \overline{\Omega}\cap D(S) \to Y$ is A-proper with respect to Γ_S for each $\lambda \in (0,1]$ with $N(\overline{\Omega}\cap D(S))$ bounded;

(b) $Sx \ne \lambda Nx + \lambda p$ for $\lambda \in (0,1)$ and $x \in \partial\Omega \cap D(S)$;

(c) $QNx + Qp \ne 0$ for $x \in S^{-1}(0) \cap \partial\Omega \cap D(S)$, where Q is the projection of Y onto Y';

(d) $Deg(S - [QN + Qp],\Omega\cap D(S),0) \ne \{0\}$.

Then there exists $x \in \overline{\Omega}\cap D(S)$ such that $Sx - Nx = p$.

Proof. Since QN is compact and $S - tN$ is A-proper with respect to Γ_S for all $t \in [0,1]$, it follows from $N(\overline{\Omega}\cap D(S))$ bounded that $H(t,x) = Sx - (1-t)[QNx + Qp] - tNx - tp$ is an A-proper homotopy with respect to Γ_S.

We may assume that $Sx - Nx \notin p$ for all $x \in \partial\Omega\cap D(S)$ (otherwise, $Sx - Nx = p$ has a solution, and we are done). We claim that $H(t,x) \ne 0$ for all $(t,x) \in [0,1]\times\partial\Omega\cap D(S)$. Indeed, if $H(t_0,x_0) = 0$ for some $(t_0,x_0) \in [0,1]\times\partial\Omega\cap D(S)$, then $t_0 \in [0,1)$.

Case (1) If $t_0 = 0$, then $Sx_0 = QNx_0 + Qp$, but $Y' \cap R(S) = \{0\}$, so $Sx_0 = 0$, which contradicts (c).

Case (2) If $t_0 \in (0,1)$, then $Sx_0 = (1-t_0)[QNx_0 + Qp] + t_0Nx + t_0p$. By (b), $QNx_0 + Qp \ne 0$, so we have

$$0 = QSx_0 = (1-t_0)Q[QNx_0 + Qp] + t_0QNx_0 + t_0Qp = QNx_0 + Qp,$$

which is a contradiction. Thus the claim is true and, as a consequence,

$$Deg(S - QN - Qp,\Omega\cap D(S),0) = Deg(S - N - p,\Omega\cap D(S),0).$$

By the assumption (d), $Sx - Nx = p$ has a solution in $\overline{\Omega} \cap D(S)$. This completes the proof.

In the following, we assume that there exists a continuous bilinear form $[\cdot, \cdot]$ on $Y \times X$ such that

$$y \in R(S) \text{ if and only if } [y, x] = 0 \text{ for all } x \in N(S). \tag{4.3.1}$$

If $\{e_1, e_2, \cdots, e_m\}$ is a basis in $N(S)$, then (4.3.1) implies that $J : R(Q) = Y' \to N(S)$ given by

$$Jy = \Sigma_{i=1}^{m}[y, e_i]e_i$$

is an isomorphism and, if $y = \Sigma_{i=1}^{m} c_i e_i$, then $[J^{-1}e_i, e_j] = \delta_{ij}$ and $[J^{-1}y, e_i] = c_i$ for $1 \leq i, j \leq m$.

For subsequent use, let $P : X \to N(S)$, $Q : Y \to Y'$ be the projections and set $A = J^{-1}P$.

Theorem 4.3.3 Let X, Y be separable Banach spaces, $S : D(S) \subseteq X \to Y$ be a Fredholm mapping of index zero and (Y_n, Q_n) be a projection scheme for Y. If Ω is an open bounded subset of X with $0 \in \Omega$ and $[\cdot, \cdot]$ is a continuous bilinear form on $Y \times X$ such that (4.3.1) holds. Also, assume that the following conditions hold:

(1) $S - \lambda N : \overline{\Omega} \cap D(S) \to Y$ is A-proper with respect to Γ_S for each $\lambda \in (0, 1]$ with $N(\overline{\Omega})$ bounded;

(2) $Sx \neq \lambda Nx + \lambda p$ for all $x \in N(S) \cap \partial\Omega \cap D(S)$ and $\lambda \in (0, 1)$;

(3) $QNx + Qp \neq 0$ for all $x \in N(S) \cap \partial\Omega$;

(4) One of the following conditions holds:

(4a) $[QNx + Qp, x] \geq 0$ for all $x \in N(S) \cap \partial\Omega$;

(4b) $[QNx + Qp, x] \leq 0$ for all $x \in N(S) \cap \partial\Omega$.

Then $Sx - Nx = p$ has a solution.

Proof. Consider the homotopy $H : [0, 1] \times \overline{\Omega} \cap D(S) \to Y$ given by

$$H(t, x) = Sx - (1 - t)(QNx + Qp) - tNx - tp$$

for all $(t, x) \in [0, 1] \times \overline{\Omega} \cap D(S)$. Since QN is compact, it follows from (1) that H is an A-proper homotopy.

Now, we claim that $H(t, x) \neq 0$ for all $(t, x) \in [0, 1] \times \partial\Omega \cap D(S)$. If this is not true, then there exists $(t_0, x_0) \in [0, 1] \times \partial\Omega \cap D(S)$ such that

$$Sx_0 = (1 - t_0)(QNx_0 + Qp) + t_0 Nx_0 + t_0 p.$$

If $t_0 = 0$, then $Sx_0 = QNx_0 + Qp$, so we have $0 = Sx_0 = QNx_0 + Qp$, which contradicts (3). If $t_0 = 1$, then $Sx_0 = Nx_0 + p$, which contradicts (2).

If $t_0 \in (0,1)$, then $Sx_0 - t_0 N x_0 - t_0 p = (1-t_0)(QNx_0 + Qp)$, so $QNx_0 + Qp \neq 0$. Therefore, we have

$$0 = QSx_0 = QNx_0 + Qp,$$

which contradicts (3). Therefore, the claim is true. Consequently, we have

$$Deg(S - QN - Qp, \Omega \cap D(S), 0) = Deg(S - N - p, \Omega \cap D(S), 0).$$

Now, we prove $Deg(S - QN - Qp, \Omega \cap D(S), 0) \neq \{0\}$. To achieve this goal, we consider the homotopy $H_1(t,x) : [0,1] \times \overline{\Omega} \cap D(S) \to Y$ given by

$$H_1(t,x) = Sx - (1-t)Ax - t(QNx + Qp)$$

for all $(t,x) \in [0,1] \times \overline{\Omega} \cap D(S)$, where $A = J^{-1}P$ if (4a) holds and $A = -J^{-1}P$ if (4b) holds.

Also, we prove that $H_1(t,x) \neq 0$ for all $(t,x) \in [0,1] \times \partial\Omega \cap D(S)$. If this is not true, then there exists $(t_0, x_0) \in [0,1] \times \partial\Omega \cap D(S)$ such that

$$Sx_0 - (1-t)Ax_0 - t_0(QNx_0 + Qp) = 0.$$

Since $N(S - A) = \{0\}$, it follows that $t_0 \neq 0$. If $t_0 = 1$, then $Sx_0 = QNx_0 + Qp = 0$, which contradicts (3). Therefore, $t_0 \in (0,1)$, so we have $Sx_0 = 0 = (1-t_0) + Ax_0 + t_0(QNx_0 + Qp)$.

Assume that (4a) holds. Since $Px = x$, we have $(1-t_0)J^{-1}x_0 + t_0(QNx_0 + Qp) = 0$ and $x_0 \neq 0$. Therefore, we have

$$(1-t_0)[J^{-1}x_0, x_0] + t_0[QNx_0 + Qp, x_0] = 0.$$

Thus $x_0 = 0$, which is impossible.

If (4b) holds, then a similar argument shows that $x_0 = 0$, which is a contradiction. Thus we have

$$Deg(S - QN - Qp, \Omega \cap D(S), 0) = Deg(S - A, \Omega \cap D(S), 0).$$

But $S - A$ is a linear injection, so $Deg(S - A, \Omega \cap D(S), 0) \neq \{0\}$. Therefore, it follows that $Deg(S - N - p, \Omega \cap D(S), 0) \neq 0$. Consequently, $Sx - Nx = p$ has a solution. This completes the proof.

Theorem 4.3.4. Let X, Y be separable Banach spaces, $S : D(S) \subseteq X \to Y$ be a Fredholm mapping of index zero, (Y_n, Q_n) be a projection scheme for Y, Ω be open bounded and symmetric about $0 \in \Omega$ and $[\cdot,\cdot]$ be a continuous bilinear form on $Y \times X$. If $N : \overline{\Omega} \cap D(S) \to Y$ is a bounded continuous mapping such that

(1) $S - \lambda N : \overline{\Omega} \cap D(S) \to Y$ is A-proper with respect to Γ_S for all $\lambda \in (0,1]$;

(2) $Sx \neq \lambda Nx + \lambda p$ for all $x \in N(S) \cap \partial\Omega \cap D(S)$ and $\lambda \in (0,1)$;

(3) $[Q(Nx+p),x][Q(N(-x)+p),x] < 0$ for all $x \in N(S) \cap \partial\Omega$.

Then $Sx - Nx = p$ has a solution in $\overline{\Omega} \cap D(S)$.

Proof. If $Sx - Nx = p$ has a solution on $\partial\Omega \cap D(S)$, we are done, so we may assume that $Sx - Nx \neq p$ for $x \in \partial\Omega \cap D(S)$.

Consider the homotopy $H : [0,1] \times \overline{\Omega} \cap D(S) \to Y$ given by

$$H(t,x) = Sx - (1-t)[Q(N(x)+p)] - tNx - tp$$

for all $(t,x) \in [0,1] \times \overline{\Omega} \cap D(S)$. Since QN is compact, it follows from (1) that $H(t,\cdot)$ is A-proper for all $t \in [0,1]$ and N is bounded. Therefore, H is an A-proper homotopy.

Now, we prove that $H(t,x) \neq 0$ for $(t,x) \in [0,1] \times \partial\Omega \cap D(S)$. In fact, assume the contrary, there exists $(t_0,x_0) \in [0,1] \times \partial\Omega \cap D(S)$ such that $H(t_0,x_0) = 0$, i.e.,

$$Sx_0 - (1-t)[Q(N(x_0)+p)] - t_0 N(x_0) - t_0 p = 0.$$

By (2), we know that $t_0 \neq 1$. If $t_0 = 0$, then $Sx_0 = Q(N(x_0)+p)$. But $R(L) \cap Y' = \{0\}$, so we have $Sx_0 = 0$ and $Q(N(x_0)+p) = 0$. Therefore, $[Q(N(x_0)+p), x_0] = 0$, which contradicts (3). Thus we must have $t_0 \in (0,1)$. Therefore,

$$0 = QSx_0 = (1-t_0)Q(N(x_0)+p) + t_0 Q(N(x_0)+p),$$

i.e., $Q(N(x_0)+p) = 0$ which contradicts (3) again. By the homotopy property of the generalized degree, we get

$$Deg(S - N - p, \Omega \cap D(S), 0) = Deg(S - Q(N+p), \Omega \cap D(S), 0).$$

Now, we prove that $\{0\} \neq Deg(S - Q(N+p), \Omega \cap D(S), 0)$. To reach this goal, we consider the homotopy $H_1(t,x) : [0,1] \times \overline{\Omega} \cap D(S) \to Y$ given by

$$H_1(t,x) = Sx - \frac{1}{1+t}[Q(N(x)+p) - t(Q(N(-x)+p)]$$

for all $(t,x) \in [0,1] \times \overline{\Omega} \cap D(S)$. Obviously, H_1 is an A-proper homotopy. We claim that $H_1(t,x) \neq 0$ for all $(t,x) \in [0,1] \times \partial\Omega \cap D(S)$. If not, there exists $(t_1,x_1) \in [0,1] \times \partial\Omega \cap D(S)$ such that $H_1(t_1,x_1) = 0$, i.e.,

$$Sx - \frac{1}{1+t_1}[Q(N(x_1)+p) - t_1(Q(N(-x_1)+p)] = 0.$$

If $t_1 = 0$, then $Sx_1 = Q(N(x_1)+p) = 0$, which contradicts (3). So $t_1 \neq 1$, thus $Lx_0 = 0$ and $Q(N(x_1)+p) - t_1(Q(N(-x_1)+p) = 0$, which lead again to a contradiction to (3). Therefore, we have

$$Deg(S - Q(N+p), \Omega \cap D(S), 0) = Deg(S - \frac{1}{2}[Q(N(\cdot)+p) - Q(N(-\cdot)+p)].$$

But $S - \frac{1}{2}[Q(N(\cdot) + p) - Q(N(-\cdot) + p)]$ is odd, so

$$0 \notin Deg(S - \frac{1}{2}[Q(N(\cdot) + p) - Q(N(-\cdot) + p)].$$

Thus we have $Deg(S - N - p, \Omega \cap D(S)) \neq \{0\}$ and, consequently, $Sx - Nx = p$ has a solution. This completes the proof.

4.4 Equations with Fredholm Mappings of Index Zero Type

In this section, we show the existence of solutions of the equations with Fredholm mappings of index zero type in Banach spaces.

Definition 4.4.1. Let X, Y be two real Banach spaces and $L : D(L) \subseteq X \to Y$ be a linear mapping. We say that L is a Fredholm mapping of index zero type if

(1) $Ker(L) = \{x \in X : Lx = 0\}$ and $Im(L) = \{Lx : x \in D(L)\}$ are closed in H;

(2) $X = Ker(L) \oplus X_1$ for some subspace X_1 of X and $Y = Y_1 \oplus Im(L)$ for some subspace Y_1 of Y;

(3) $Ker(L)$ is linearly homeomorphic to $Coker(L) = Y/Im(L)$.

Remark. Obviously, if X is linearly homeomorphic to Y, $L = 0$ is a Fredholm mapping of index zero type but not a Fredholm mapping of index zero. If L is a Fredholm mapping of index zero, then

$$dim(Ker(L)) = dim(Coker(L)) < +\infty,$$

so $Ker(L)$ is linearly homeomorphic to $Coker(L)$ and thus L is a Fredholm mapping of index zero type.

Now, assume that $L : D(L) \subset X \to Y$ is a Fredholm mapping of index zero type. Then there exist linear projections $P : X \to X$ and $Q : Y \to Y$ such that

$$Im(P) = Ker(L), \quad Im(Q) = Y_1.$$

Obviously, the restriction of L_P of L to $D(L) \cap Ker(P)$ is one to one and onto $Im(L)$, so its inverse $K_P : Im(L) \to D(L) \cap Ker(P)$ is defined. Let $J : Ker(L) \to Y_1$ be a linear homeomorphism and set $K_{PQ} = K_P(I - Q)$

Proposition 4.4.2. $L + \lambda JP : X \to Y$ is a bijective mapping for each $\lambda \neq 0$.

Proof. For each $\lambda \neq 0$, if $Lx + \lambda JPx = 0$, then $JPx = 0$ and $Lx = 0$, so $x \in Ker(L)$. Thus $x = 0$. On the other hand, for $y = y_1 + y_2 \in Y$, $y_1 \in Y_1, y_2 \in Im(L)$, put $x = \lambda^{-1}J^{-1}y_1 + K_P y_2$, then $Lx + \lambda JPx = y$. Therefore, $L + \lambda JP$ is bijective.

Proposition 4.4.3. Let X, Y be real separable Banach spaces, (Y_n, Q_n) be a projection scheme for Y and $L : D(L) \subset X \to Y$ be a Fredholm mapping of zero index type. Then, for each $\lambda \neq 0$, there exists a projection scheme $\Gamma_{\lambda,L}$ for (X, Y).

Proof. For each $\lambda \neq 0$, put $K_\lambda = L + \lambda JP$. By Proposition 4.4.2, K_λ is bijective. Set $X_n = K_\lambda^{-1} Y_n$ for $n = 1, 2, \cdots$. Obviously, we have

$$dim(X_n) = dim(Y_n), \quad X = \overline{\cup_{n=1}^{\infty} X_n}.$$

Thus $\Gamma_{\lambda,L} = \{X_n, Y_n, Q_n\}$ is a projection scheme for (X, Y).

Proposition 4.4.4. Let $L : D(L) \subset X \to Y$ be a Fredholm mapping of zero index type. Assume that X is reflexive. If $G \subset X$ is bounded closed and convex, then $L : G \cap D(L) \to Y$ is pseudo A-proper with respect to $\Gamma_{\lambda,L}$ for each $\lambda \neq 0$.

Proof. For any sequence (x_{n_k}) in $G \cap D(L) \cap X_{n_k}$ with $Q_{n_k} L x_{n_k} \to y$, we may assume that $x_{n_k} \rightharpoonup x_0 \in G$ by taking subsequences. Notice that

$$Q_{n_k}(Lx_{n_k} + \lambda JPx_{n_k}) = Lx_{n_k} + \lambda JPx_{n_k}, \quad JPx_{n_k} \rightharpoonup JPx_0,$$

so we have

$$x_{n_k} = (L + JP)^{-1}(Q_{n_k}(Lx_{n_k} + JPx_{n_k}) \rightharpoonup (L + JP)^{-1}(y + JPx_0) = x_0.$$

Thus $x_0 \in D(L)$ and $Lx_0 = y$. Therefore, L is pseudo A-proper with respect to $\Gamma_{\lambda,L}$.

Definition 4.4.5. Let X be a real separable Banach space, $\Gamma_0 = (X_n, P_n)$ be a projection scheme for X, Y be a real Banach space, $L : D(L) \subset X \to Y$ be a Fredholm mapping of zero index type and $N : D \subset X \to Y$ be a mapping.

(1) If $I - P - (J^{-1}Q + K_{PQ})N$ is A-proper with respect to Γ_0, then we say that N is L-A-proper with respect to Γ_0;

(2) If $I - P - (J^{-1}Q + K_{PQ})N$ is pseudo A-proper with respect to Γ_0, then we say that N is pseudo L-A-proper with respect to Γ_0;

(3) A family of mappings $H(t, x) : [0, 1] \times D \to Y$ is called a homotopy of L-A-proper mappings with respect to Γ_0 if $H(t, \cdot)$ is a L-A-proper mapping with respect to Γ_0 for all $t \in [0, 1]$.

Proposition 4.4.6. Let $L : D(L) \subseteq X \to Y$ be a linear mapping with $Ker(L) = \{0\}$ and $Im(L) = Y$, then the following conclusions hold:

(1) If $\Gamma_0 = (X_n, P_n)$ is a projection scheme for X, then 0 is L-A-proper with respect to Γ_0;

(2) If (Y_n, Q_n) is a projection scheme for Y and L^{-1} is continuous, then L is A-proper with respect to $\Gamma_{1,L}$, where $\Gamma_{1,L}$ is constructed as in Proposition 4.4.3.

Proof. (1) We have $P = 0$, $Q = 0$ and the identity mapping $I : X \to X$ is obviously A-proper with respect to Γ_0. Thus 0 is $L-$A-proper with respect to Γ_0.

(2) Since $Ker(L) = \{0\}$, the mapping K in the proof of Proposition 4.4.3 is just the mapping L. Thus $X_n = L^{-1}Y_n$. If $x_{n_k} \in X_{n_k}$ such that $Q_{n_k}Lx_{n_k} \to y$, then $Lx_{n_k} = Q_{n_k}Lx_{n_k} \to y$. Therefore, we have $x_{n_k} \to L^{-1}y$. The conclusion is true.

Proposition 4.4.7. Let $L : D(L) \subset X \to Y$ be a Fredholm mapping of zero index type and $\Gamma_0 = (X_n, P_n)$ be a projection scheme for X. If $G \subset X$ is a bounded closed convex subset, $T : G \to Y$ is a weakly continuous mapping and X is reflexive, then T is pseudo L-A-proper with respect to Γ_0.

Proof. For any subsequence (x_{n_k}) in X_{n_k} such that $P_{n_k}(I - P - J^{-1}QT - K_{PQ}T)x_{n_k} \to y$, we may assume that $x_{n_k} \rightharpoonup x_0 \in G$ by taking a subsequence. Thus we have

$$(I - P)x_{n_k} \rightharpoonup x_0, \quad J^{-1}QTx_{n_k} \rightharpoonup J^{-1}QTx_0$$

and

$$K_{PQ}Tx_{n_k} \rightharpoonup K_{PQ}Tx_0$$

and, consequently, $(I - P - J^{-1}QT - K_{PQ}T)x_0 = y$. So T is pseudo $L-$A-proper with respect to Γ_0.

Proposition 4.4.8. Let X, Y be real separable Banach spaces, (Y_n, Q_n) be a projection scheme for Y. Let $L : D(L) \subset X \to Y$ be a Fredholm mapping of zero index type, $G \subset X$ be a bounded closed subset and $N : G \to Y$ be a continuous compact mapping. Then $L + \lambda JP - N$ is A-proper with respect $\Gamma_{\lambda,L}$ for each $\lambda > 0$.

Proof. For any sequence (x_{n_k}) in $G \cap D(L) \cap X_{n_k}$ with $Q_{n_k}(L + \lambda JP - N)x_{n_k} \to y$, in view of the compactness of N, we may assume that $Nx_{n_k} \to y_0 \in Y$ by taking a subsequence. Notice that

$$Q_{n_k}(Lx_{n_k} + \lambda JPx_{n_k}) = Lx_{n_k} + \lambda JPx_{n_k},$$

so we have

$$\begin{aligned} x_{n_k} &= (L + JP)^{-1}[Q_{n_k}(L + \lambda JP - N)x_{n_k} + Q_{n_k}Nx_{n_k}] \\ &\to (L + \lambda JP)^{-1}(y + y_0) = x_0. \end{aligned}$$

Thus $x_0 \in D(L)$, $Nx_0 = y_0$ and $(L+\lambda JP-N)x_0 = y$. Therefore, $L+\lambda JP-N$ is A-proper with respect to $\Gamma_{\lambda,L}$.

In the following, suppose that X, Y are real separable Banach spaces, $L : D(L) \subseteq X \to Y$ is a Fredholm mapping of index zero type with $D(L)$ dense in X and $N : \overline{\Omega} \subset X \to Y$ is a nonlinear mapping. We consider the semilinear operator equation $Lx - Nx = 0$ and we apply the generalized degree theory in Section 4.2 to study such an equation.

Lemma 4.4.9. Let $L : D(L) \subseteq X \to Y$ be a Fredholm mapping of index zero type, $\Omega \subset X$ an open bounded subset and $N : \overline{\Omega} \to Y$ be a mapping. If $0 \notin (L-N)(\partial\Omega \cap D(L))$, then $0 \notin [I - P - (J^{-1}Q + K_{PQ})N](\partial\Omega)$.

Proof. Suppose the contrary. Then there exists $x_0 \in \partial\Omega$ such that $0 \in x_0 - Px_0 - (J^{-1}Q + K_{PQ})Nx_0$. Since $J^{-1}QTx_0 \in Ker(L) = Im(P)$, $x_0 - Px_0 \in Ker(P)$ and $K_{pQ}Tx_0 \in D(L) \cap Ker(P)$, we must have

$$J^{-1}QNx_0 = 0, x_0 - Px_0 - K_{PQ}Nx_0 = 0.$$

Therefore, we have

$$QNx_0 = 0, x_0 - Px_0 - K_PNx_0 = 0, \text{ i.e., } Lx_0 - Nx_0 = 0,$$

which contradicts $0 \notin (L-N)(\partial\Omega \cap D(L))$. This completes the proof.

Now, let $L : D(L) \subseteq X \to Y$ be a Fredholm mapping of index zero type, $\Gamma_0 = (X_n, P_n)$ be a projection scheme for X, $\Omega \subset X$ be an open bounded subset and $N : \overline{\Omega} \to Y$ be a L-A-proper mapping respect to Γ_0. Suppose that $0 \notin (L-T)(\partial\Omega \cap D(L))$. By Lemma 4.4.9, we have

$$0 \notin [I - P - (J^{-1}Q + K_{PQ})N](\partial\Omega).$$

Since $I - P - (J^{-1}Q + K_{PQ})N$ is an A-proper mapping with respect to Γ_0, the generalized degree $deg(I - P - (J^{-1}Q + K_{PQ})N, \Omega, 0)$ is well defined and we define

$$deg_{\Gamma_0,J}(L - N, \Omega, 0) = deg(I - P - (J^{-1}Q + K_{PQ})N, \Omega, 0), \qquad (4.4.1)$$

which is called the generalized coincidence degree of L and N on Ω.

Theorem 4.4.10. The generalized coincidence degree of L and N defined by (4.4.1) on Ω has the following properties:

(1) If Ω_1 and Ω_2 are disjoint open subsets of Ω such that $0 \notin (L-N)(D(L) \cap \overline{\Omega \setminus (\Omega_1 \cup \Omega_2)})$, then

$$deg_{\Gamma_0,J}(L - N, \Omega, 0) \subseteq deg_{\Gamma_0,J}(L - N, \Omega_1) + deg_{\Gamma_0,J}(L - N, \Omega_2, 0);$$

(2) If $H(t,x) : [0,1] \times \overline{\Omega} \to Y$ is a homotopy of L-A-proper mappings with respect to Γ_0 and $0 \neq Lx - H(t,x)$ for all $(t,x) \in [0,1] \times \partial\Omega \cap D(L)$, then $deg_{\Gamma_0,J}(L - H(t,\cdot), \Omega, 0)$ doesn't depend on $t \in [0,1]$;

(3) If $deg_{\Gamma_0}(L-N,\Omega,0)\neq\{0\}$, then $0\in(L-N)(D(L)\cap\Omega)$;

(4) If $L:D(L)\subseteq X\to Y$ is a linear mapping such that $L^{-1}:Y\to D(L)$ is continuous, then $deg_{\Gamma_0,J}(L,\Omega,0)=\{1\}$ if $0\in\Omega$;

(5) If Ω is a symmetric neighborhood of 0 and $N:\overline{\Omega}\to Y$ is an odd L-A-proper mapping respect to Γ_0 with $0\notin(L-N)(\partial\Omega\cap D(L))$, then $deg_{\Gamma_0,J}(L-N,\Omega,0)$ does not contain even numbers.

Proof. (1)-(3) follows directly from the definition and the properties of generalized degree.

(4) Since $Ker(L)=\{0\}$, $P=0$ and $Q=0$, the zero mapping is $L-$A-proper with respect to Γ_0. Thus $deg_{\Gamma_0,J}(L,\Omega,0)=deg(I,\Omega,0)=\{1\}$.

(5) Since N is odd, the mapping $I-P-(J^{-1}Q+K_{PQ})N$ is odd and thus $deg(I-P-(J^{-1}Q+K_{PQ})N,\Omega,0)$ doesn't contain even numbers. The conclusion follows by definition.

Corollary 4.4.11. Let $L:D(L)\subseteq X\to Y$ be a linear mapping such that $L^{-1}:Y\to D(L)$ is continuous, $\Omega\subset X$ be an open bounded subset with $0\in\Omega$ and $N:\overline{\Omega}\to Y$ be a mapping such that $\{L-tN\}_{t\in[0,1]}$ is a homotopy of L-A-proper mappings respect to Γ_0. If $Lx\notin tNx$ for all $(t,x)\in[0,1]\times\partial\Omega\cap D(L)$, then $deg(L-N,\Omega,0)=1$.

In the following, let $L:D(L)\subset X\to Y$ be a densely defined Fredholm mapping of zero index type. We assume that $\Gamma_0=(Y_n,Q_n)$ is a projection scheme for Y, $\Gamma_{\lambda,L}$ is the same as in Proposition 4.4.3 and $L+\lambda JP-N$ is A-proper with respect to $\Gamma_{\lambda,L}$ for $\lambda\in(0,\lambda_0)$, where $\lambda_0>0$ is a constant. Suppose that $0\notin\overline{(L-N)(D(L)\cap\partial\Omega)}$. Then there exists $\lambda_1<\lambda_0$ such that

$$0\notin(L+\lambda JP-N)(D(L)\cap\partial\Omega)\quad\text{for all }\lambda\in(0,\lambda_1).$$

We define a generalized degree by

$$deg(L-N,\Omega,0)=\cap_{0<\lambda<\lambda_1}\cup_{0<\epsilon\leq\lambda}deg(L+\epsilon JP-N,\Omega,0),\qquad(4.4.2)$$

where $deg(L+\epsilon JP-N,\Omega,0)$ is the generalized degree for A-proper mappings.

Notice that, if $0\notin(L+\lambda JP-N)(D(L)\cap\partial\Omega)$ for all $\lambda\in(0,\lambda_2)$, then it is easy to check that

$$\begin{aligned}&\cap_{0<\lambda<\lambda_1}\cup_{0<\epsilon\leq\lambda}deg(L+\epsilon JP-N,\Omega,0)\\&=\cap_{0<\lambda<\lambda_2}\cup_{0<\epsilon\leq\lambda}deg(L+\epsilon JP-N,\Omega,0).\end{aligned}$$

Thus (4.4.2) is well defined.

Theorem 4.4.12. The generalized degree defined by (4.4.2) has the following properties:

(1) If Ω_1 and Ω_2 are two open subsets of Ω such that $\Omega_1 \cap \Omega_2 = \emptyset$ and $0 \notin \overline{(L-N)(D(L) \cap \overline{\Omega} \setminus (\Omega_1 \cup \Omega_2))}$, then

$$deg(L-N, \Omega, 0) \subseteq deg(L-N, \Omega_1) + deg(L-N, \Omega_2, 0);$$

(2) If $H(t,x) : [0,1] \times \overline{\Omega} \to Y$ satisfies $0 \notin \overline{\cup_{t \in [0,1]} (L - H(t, \cdot)(D(L) \cap \partial\Omega)}$ and $\{L + \lambda JP - H(t, \cdot)\}_{t \in [0,1]}$ is a homotopy of A-proper mappings with respect to $\Gamma_{\lambda,L}$ for each $\lambda \in (0, \lambda_0)$, where $\lambda_0 > 0$ is a constant, then $deg(L - H(t, \cdot), \Omega, 0)$ does not depend on $t \in [0,1]$;

(3) If $deg_{\Gamma_0}(L-N, \Omega, 0) \neq \{0\}$, then $0 \in \overline{(L-N)(D(L) \cap \Omega)}$;

(4) If Ω is a symmetric neighborhood of 0 and $N : \overline{\Omega} \to Y$ is an odd mapping such that $L + \lambda JP - N$ is A-proper with respect to $\Gamma_{\lambda,L}$ for all $\lambda > 0$, and $0 \notin \overline{(L-N)(\partial\Omega \cap D(L))}$, then $deg(L-N, \Omega, 0)$ does not contain even numbers;

(5) $deg(L, \Omega, 0) \subseteq \{\pm 1\}$ if $0 \in \Omega$.

Proof. (1) By assumption, there exists $\lambda_0 > 0$ such that

$$0 \notin (L + \lambda JP - N)(D(L) \cap \overline{\Omega \setminus (\Omega_1 \cup \Omega_2)})$$

for all $\lambda \in (0, \lambda_0)$. If $m \in deg(L-N, \Omega, 0)$, then there exist $\lambda_j \to 0^+$, $\lambda_j < \lambda_0$, $j = 1, 2, \cdots$, such that $m \in deg(L + \lambda_j JP - N, \Omega, 0)$. By Theorem 4.2.3, we have

$$\begin{aligned} &deg(L + \lambda_j JP - N, \Omega, 0) \\ &\subseteq deg(L + \lambda_j JP - N, \Omega_1, 0) + deg(L + \lambda_j JP - N, \Omega_2, 0) \end{aligned}$$

for $j = 1, 2, \cdots$. By (4.4.2), (1) is true.

(2) Since $0 \notin \overline{\cup_{t \in [0,1]} (L - H(t, \cdot)(D(L) \cap \partial\Omega)}$, there exists $\lambda_1 > 0$ such that

$$0 \notin \cup_{t \in [0,1]} (L + \lambda JP - H(t, \cdot))(\partial\Omega \cap D(L)) \quad \text{for all } \lambda \in (0, \lambda_1).$$

By Theorem 4.2.3, $deg(L + \lambda JP - H(t, \cdot), \Omega, 0)$ does not depend on $t \in [0,1]$ for $\lambda \in (0, \min\{\lambda_0, \lambda_1\})$. So the conclusion of (2) follows from (4.4.2).

(3) If $deg_{\Gamma_0}(L-N, \Omega, 0) \neq \{0\}$, then there exists $0 \neq m \in deg_{\Gamma_0}(L-N, \Omega, 0)$, so there exists $\lambda_j \to 0^+$ such that $m \in deg(L + \lambda_j JP - N, \Omega, 0)$. Therefore, $(L + \lambda_j JP - N)x$ has a solution in $\Omega \cap D(L)$ for $j = 1, 2, \cdots$. By letting $j \to \infty$, we get $0 \in \overline{(L-N)(D(L) \cap \Omega)}$.

(4) The proof is left to the reader.

(5) Now, $L + \lambda JP$ is A-proper with respect to $\Gamma_{\lambda,L}$ and $0 \notin (L + \lambda JP)(\partial\Omega \cap D(L))$ for all $\lambda > 0$. Since $L + \lambda JP)$ is bijective, $deg(L + \lambda JP, \Omega, 0) \subseteq \{\pm 1\}$ for all $\lambda > 0$, so we have

$$deg(L-N, \Omega, 0) \subseteq \{\pm 1\}.$$

This completes the proof.

Theorem 4.4.13. Let X, Y be real separable Banach spaces and (Y_n, Q_n) be a projection scheme for Y. Let $L : D(L) \subset X \to Y$ be a Fredholm mapping of zero index type, $0 \in \Omega \subset X$ be a bounded subset and $N : \overline{\Omega} \to Y$ be a continuous compact mapping. Suppose that the following conditions are satisfied:

(1) $0 \notin \overline{(L-N)(\partial\Omega \cap D(L))}$;

(2) $0 \notin \overline{QN(\partial\Omega \cap D(L))}$.

Then $deg(L-N, \Omega, 0) = deg(L - QN, \Omega, 0)$.

Proof. For each $\lambda \in (0, \lambda_0)$, a similar proof to Proposition 4.4.3 shows that $\{L + \lambda JP - tN - (1-t)QN\}_{t\in[0,1]}$ is a homotopy of A-proper mappings with respect to $\Gamma_{\lambda,L}$.

Now, we claim that

$$0 \notin \overline{\cup_{t\in[0,1]}(L - tN - (1-t)QN)(D(L) \cap \partial\Omega)}.$$

If this is not true, then there exist $t_j \in [0,1]$ with $t_j \to t_0$ and $x_j \in \partial\Omega \cap D(L)$ such that $Lx_j - t_j Nx_j - (1-t_j)QNx_j \to 0$.

Case (1) If $t_0 = 1$, then $Lx_j - Nx_j \to 0$, which contradicts the assumption (1).

Case (2) If $t_0 \neq 1$, then $QLx_j - QNx_j \to 0$ and thus we have $QNx_j \to 0$ and $x_j \in D(L)$, which contradicts the assumption (2).

By (2) of Theorem 4.4.12, we get $deg(L-N, \Omega, 0) = deg(L - QN, \Omega, 0)$.

Finally, let $L : D(L) \subseteq X \to Y$ be a Fredholm mapping of index zero type, $\Gamma_0 = (X_n, P_n)$ be a projection scheme for X and $\Omega \subset X$ an open bounded subset. Let $N : \overline{\Omega} \to Y$ be a mapping such that $I - (L + \lambda JP)^{-1}(N + \lambda JP)$ is A-proper with respect Γ_0 for some $\lambda > 0$. One can easily see that

$$0 \in Lx - Nx \quad \text{if and only if} \quad 0 \in (I - (L + \lambda JP)^{-1}(N + \lambda JP))x.$$

Assume that $0 \notin (L - N)(\partial\Omega \cap D(L))$. Then we have

$$0 \notin (I - (L + \lambda JP)^{-1}(N + \lambda JP))(\partial\Omega) \quad \text{for all } \lambda > 0$$

and we define a generalized degree by

$$deg_{\Gamma_0}(L - N, \Omega, 0) = \cup_{0<\lambda} deg(I - (L + \lambda JP)^{-1}(N + \lambda JP), \Omega, 0), \quad (4.4.3)$$

where $deg(I - (L + \lambda JP)^{-1}(N + \lambda JP), \Omega, 0)$ is the generalized degree for A-proper mappings if $I - (L + \lambda JP)^{-1}(N + \lambda JP)$ is A-proper. Otherwise, we have

$$deg(I - (L + \lambda JP)^{-1}(N + \lambda JP), \Omega, 0) = \emptyset.$$

Theorem 4.4.14. The generalized degree defined by (4.4.3) has the following properties:

(1) If Ω_1 and Ω_2 are disjoint open subsets of Ω such that $0 \notin (L-N)(D(L) \cap \overline{\Omega \setminus (\Omega_1 \cup \Omega_2)})$, then

$$deg(L-N, \Omega, 0) \subseteq deg(L-N, \Omega_1) + deg(L-N, \Omega_2, 0);$$

(2) If $H(t,x) : [0,1] \times \overline{\Omega} \to Y$ satisfies $0 \notin \cup_{t\in[0,1]}(L - H(t,\cdot)(D(L) \cap \partial\Omega)$ and $\{I - (L+\lambda JP)^{-1}(H(t,\cdot) + \lambda JP)\}_{t\in[0,1]}$ is a homotopy of A-proper mappings with respect to $\Gamma_{\lambda,L}$ for all $\lambda > 0$, then $deg(L - H(t,\cdot), \Omega, 0)$ does not depend on $t \in [0,1]$;

(3) If $deg(L-N, \Omega, 0) \neq \{0\}$, then $0 \in (L-N)(D(L) \cap \Omega)$;

(4) If Ω is a symmetric neighborhood of 0 and $N : \overline{\Omega} \to Y$ is an odd mapping such that $I - (L+\lambda JP)^{-1}(N + \lambda JP)$ is A-proper with respect to Γ_0 for some $\lambda > 0$ and $0 \notin (L-N)(\partial\Omega \cap D(L))$, then $deg(L-N, \Omega, 0)$ does not contain even numbers.

Proof. The proof is standard. We prove (2) and skip the others. Since $0 \notin \cup_{t\in[0,1]}(L - H(t,\cdot)(D(L) \cap \partial\Omega)$, it follows that

$$0 \notin \cup_{t\in[0,1]}(I - (L+\lambda JP)^{-1}(H(t,\cdot) + \lambda JP)(\partial\Omega) \quad \text{for all } \lambda > 0.$$

By Theorem 4.2.3, we know that

$$deg(I - (L+\lambda JP)^{-1}(H(t,\cdot) + \lambda JP), \Omega, 0)$$

does not depend on $t \in [0,1]$ for each $\lambda > 0$. Thus the conclusion of (2) follows from (4.4.3). This completes the proof.

Theorem 4.4.15. Suppose that $(L+\lambda JP)^{-1} : Y \to X$ is a continuous compact mapping for each $\lambda > 0$, $\Omega \subset X$ is an open bounded subset with $0 \in \Omega$ and $N : \overline{\Omega} \to Y$ is a continuous bounded mapping such that $Lx \neq Nx$ and $QNx \neq \eta JPx$ for all $x \in \partial\Omega \cap D(L)$ and $\eta > 0$, where P, Q are projections as in the beginning of this section. Then $deg(L-N, \Omega, 0) = \{1\}$.

Proof. Since $(L+\lambda JP)^{-1} : Y \to X$ is continuous and compact for each $\lambda > 0$, it follows that $\{I - (L+\lambda JP)^{-1}t(N+\lambda JP)\}_{t\in[0,1]}$ is a homotopy of A-proper mappings.

Now, we claim that

$$x \neq (L+\lambda JP)^{-1}t(N+\lambda JP)x$$

for all $(t,x) \in [0,1] \times (\partial\Omega \cap D(L))$ and $\lambda > 0$. If this is not true, then there exist $\lambda_0 > 0$ and $(t_0, x_0) \in [0,1) \times \partial\Omega$ such that $x_0 = (L+\lambda_0 JP)^{-1}t_0(Nx_0 + \lambda JPx_0)$. Thus we have $x_0 \in D(L)$ and

$$Lx_0 + \lambda_0 JPx_0 = t_0(Nx_0 + \lambda_0 JPx_0).$$

Obviously, $t_0 \neq 1$ and so $(1-t_0)\lambda_0 JPx_0 = t_0 QNx_0$, which contradicts our assumption. Consequently, the A-proper degree

$$deg(I-(L+\lambda JP)^{-1}(N+\lambda JP),\Omega,0) = deg(I,\Omega,0) = \{1\}.$$

Therefore, from (4.4.3), we get

$$deg(L-T,\Omega,0) = \{1\}.$$

This completes the proof.

Corollary 4.4.16. Suppose that H is a separable Hilbert space, $(L+\lambda JP)^{-1} : H \to X$ is a continuous compact mapping for each $\lambda > 0$ and $\Omega \subset X$ is an open bounded subset $0 \in \Omega$, $N : \overline{\Omega} \to H$ is a continuous bounded mapping such that $Lx \neq Nx$ for all $x \in \partial\Omega \cap D(L)$, $QNx \neq 0$ for $x \in \partial\Omega \cap D(L) \cap Ker(P)$ and $(QNx, JPx) < 0$ for all $x \in \partial\Omega \cap D(L) \cap (Ker(P))^c$, where P, Q are projections as in the begining of this section. Then $deg(L-N,\Omega,0) = \{1\}$.

Proof. From our assumption, we have $QNx \neq \eta JPx$ for all $x \in \partial\Omega \cap D(L)$ and $\eta > 0$. Thus the conclusion follows from Theorem 4.4.15.

4.5 Applications of the Generalized Degree

In this section, we apply the results in Sections 4.3 and 4.4 to the periodic semilinear ordinary and partial differential equations.

First, consider the periodic ordinary differential equations:

$$\begin{cases} x''(t) = f(t,x(t),x'(t),x''(t)) - g(t), & t \in [0,T], \\ x(0) = x(T),\ x'(0) = x'(T) \end{cases} \qquad (E\ 4.5.1)$$

where $f : [0,T] \times R^3 \to R$ is a continuous function.

Lemma 4.5.1. Let $f : [0,T] \times R^3 \to R$ be a continuous function satisfying the following:

(a) there exist constants $M > 0$ and $c, d \in R$ with $c \leq g_m \leq g_M \leq d$, and $x \leq M$ implies that $d < f(t,x,0,r)$ for all $t, \in [0,T]$ and $r \in R^3$, while $x \leq -M$ implies that $f(t,x,0,r) < c$ for all $t \in [0,T]$ and $r \in R^3$, where $g_m = \min\{g(t) : t \in [0,T]\}$ and $g_M = \max\{g(t) : t \in [0,T]\}$.

If $x(t)$ is a C^2 solution of $(E\ 4.5.1)$ and $|x(t)|$ does not achieve its maximum at $t=0$ or $t=T$, then

$$|x(t)| \leq M \quad \text{for all} \quad t \in [0,T].$$

Proof. Assume that $|x(t)|$ achieves its maximum at $t_0 \in (0, T)$. We claim that $|x(t_0)| \leq M$. If not, then $|x(t_0)| > M$.

Case (1) If $x(t_0) > M$, then $x'(t_0) = 0$ and $x''(t_0) \leq 0$. Therefore, we have

$$f(t_0, x(t_0), 0, x''(t_0)) - g(t_0) \leq 0.$$

By (a), we have $d - g(t_0) < 0$, so $g(t_0) < d$, which contradicts $g_M \leq d$.

Case (2) If $x(t_0) < -M$, then $x'(t_0) = 0$ and $x''(t_0) \geq 0$. Therefore, we have

$$f(t_0, x(t_0), 0, x''(t_0)) - g(t_0) \geq 0.$$

By (a), we have $b - g(t_0) > 0$, so $c > g(t_0)$, which contradicts $g_m > c$.

Therefore, from the above arguments, we know that $|x(t_0)| \leq M$ and Lemma 4.5.1 is proved. This completes the proof.

Lemma 4.5.2. Suppose that the condition (a) of Lemma 4.5.1 holds. Then any solution $x(\cdot)$ of the problem $(E\ 4.5.1)$ satisfies

$$|x(t)| \leq M \quad \text{for all} \quad t \in [0, T].$$

Proof. Let $x(\cdot)$ be a solution of $(E\ 4.5.1)$. Assume that $|x(t)|$ achieves its maximum at $t = 0$. Then $x'(0) = 0$. Otherwise, $|x(t)|$ can not achieve its maximum at $t = 0$. Therefore, $|x(0)| = |x(T)| \leq M$.

Lemma 4.5.3. Suppose the following conditions hold:

(1) There exists a constant $M > 0$ such that, for each solution $x(\cdot)$ of $(E$ $4.4.1)$, $|x(t)| \leq M$ for $t \in [0, T]$;

(2) There exist constants $A, C > 0$ and $B \in [0, 1]$ such that

$$|f(t, x, r, q)| \leq Ar^2 + B|q| + C$$

for all $(t, x) \in [0, T] \times [-M, M]$ and $r, q \in R$.

Then there are constants $M_1, M_2 > 0$ depending only on M, A, B, C and g_M such that

$$|x'(t)| \leq M_1, \quad |x''(t)| \leq M_2 \quad \text{for all } t \in [0, T]$$

for each solution $x(\cdot)$ of $(E\ 4.5.1)$.

Proof. Since $x'(t)$ vanishes at least once in $[0, T]$, each point $t \in [0, T]$ for which $x'(t) \neq 0$ belongs to an interval $[\mu, \gamma]$ such that $x'(t)$ maintains a fixed sign on $[\mu, \gamma]$ and $x(\mu)$ or $x(\gamma)$ is 0. Without loss of generality, we may assume that $x(\mu) = 0$ and $x'(t) \geq 0$ for $t \in [\mu, \gamma]$. Therefore, it follows from (2) that

$$|x''(t)| \leq Ax'(t)^2 + B|x''(t)| + C + D, \tag{4.5.1}$$

where $D = \max\{|g_m|, |g_M|\}$. Multiply (4.5.1) by $x'(t)$ and rearrange the terms, we obtain

$$(1 - B)|x''(t)x'(t)| \leq (Ax'(t)^2 + C + D)x'(t).$$

Set $\alpha = \frac{A}{1-B}$ and $\beta = \frac{C+D}{1-B}$, we get

$$\frac{2\alpha x''(t)x'(t)}{\alpha x'(t)^2 + \beta} \leq 2\alpha x'(t).$$

Integrate the last inequality over $[\mu, t]$ and use $|x(t)| \leq M$ and $x'(\mu) = 0$, we obtain

$$ln(\frac{\alpha x'(t)^2 + \beta}{\beta}) \leq 2\alpha M.$$

Therefore, we have

$$|x'(t)| \leq [\frac{\beta}{\alpha}(e^{2\alpha M} - 1)]^{\frac{1}{2}} = M_1.$$

This and the condition (2) implies that $|x''(t)| \leq M_2$ for some M_2 depending on M, A, B, and C. This completes the proof.

Combine Lemma 4.5.1 and Lemma 4.5.2 with Lemma 4.5.3, we get the following:

Proposition 4.5.4. Assume that the condition of Lemma 4.4.1 holds. If there are continuous functions $A(t, x), C(t, x) > 0$ which are bounded on compact subsets of $[0, T] \times R$ and a constant $B \in [0, 1]$ such that

$$|f(t, x, r, q)| \leq A(t, x)r^2 + B|q| + C(t, x)$$

for all $(t, x) \in [0, T] \times [-M, M]$ and $r, q \in R$, then there are constants M_1 and M_2 such that, for any solution of $(E\ 4.5.1)$,

$$|x(t)| \leq M, \ |x'(t)| \leq M_1, \ |x''(t)| \leq M_2 \text{ for all } t \in [0, T].$$

Now, we consider a family of the periodic problems:

$$\begin{cases} x''(t) = \lambda f(t, x(t), x'(t), x''(t)) - \lambda g(t), & t \in [0, T], \\ x(0) = x(T), \ x'(0) = x'(T), \end{cases} \qquad (E\ 4.5.2)$$

where $\lambda \in [0, 1]$.

Lemma 4.5.5. Suppose that the following conditions hold:

(1) Let $M > 0$ and c, d be the same as in Lemma 4.5.1 and the condition of Lemma 4.5.1 holds;

(2) Let $A, C > 0$ and $B \in [0,1]$ be such that

$$|f(t,x,r,q)| \leq Ar^2 + B|q| + C$$

for all $(t,x) \in [0,T] \times [-M,M]$ and $r,q \in R$.

Then there are constants $M_1, M_2 > 0$ such that, for $\lambda \in [0,1]$ and any solution $x_\lambda(\cdot)$ of $(E\ 4.5.2)$,

$$|x_\lambda(t)| \leq M, \ |x'_\lambda(t)| \leq M_1, \ |x''_\lambda(t)| \leq M_2 \text{ for all } t \in [0,T].$$

Proof. Replace f by λf and g by λg in Lemma 4.5.2 and Lemma 4.5.3, respectively, one easily gets the conclusions.

To formulate an existence result for the problem $(E\ 4.5.1)$, let $Y = C[0,T]$ be the Banach space of continuous functions on $[0,T]$ with the supremum norm and $C^k([0,T])$ be the Banach space of k-times continuously differentiable functions with the norm $\|x(\cdot)\|_k = \max\{\|x^{(i)}(\cdot)\| : 0 \leq i \leq k\}$, where $\|\cdot\|$ is the norm in $C([0,T])$. Set

$$X = \{x(\cdot) \in C^2([0,T]) : x(0) = x(T), \ x'(0) = x'(T)\},$$

and let $S : X \to Y$ be a mapping defined by

$$Sx(t) = x''(t) \quad \text{for all } x(\cdot) \in X,\ t \in [0,T].$$

It is well known that S is a Fredholm mapping of index zero, $N(S) = \{x(\cdot) \in X : x(t) \text{ is constant}\}$, $R(S) = \{y(\cdot) \in Y : \int_0^T y(t)dt = 0\}$, $X = N(S) \oplus X'$ and $Y = N(S) \oplus R(S)$. It is easy to see that $K = S - I : X \to Y$ is a linear isomorphism. Let (Y_n, Q_n) be a projection scheme for Y and set $X_n = K^{-1}Y_n$, then $d(x, X_n) \to 0$ as $n \to \infty$, so $\Pi_S = (X_n, Y_n, Q_n)$ is admissible for mappings from X to Y. Now, S is A-proper with respect to Π_S.

Finally, we set $N : X \to Y$ by

$$Nx(t) = f(t, x(t), x'(t), x''(t)) \quad \text{for all } t \in [0,T],\ x(\cdot) \in X.$$

Then N is continuous and maps bounded subsets in X to bounded subsets in Y.

Now, we have the following:

Theorem 4.5.6. Let $g(\cdot) \in Y$, $f(t,x,r,q) : [0,T] \times R^3 \to R$ be a continuous function and S, Π_S, N be the same as the above. Suppose that the following conditions hold:

(1) $S - \lambda N$ is A-proper with respect to Π_S for each $\lambda \in (0,1]$;

(2) There exist constants $M > 0$ and $c, d \in R$ with $c \leq g_m \leq g_M \leq d$, and $x \leq M$ implies that $d < f(t, x, 0, r)$ for all $t \in [0, T]$ and $r \in R^3$, while $x \leq -M$ implies that $f(t, x, 0, r) < c$ for all $t \in [0, T]$ and $r \in R^3$, where $g_m = \min\{g(t) : t \in [0, T]\}$ and $g_M = \max\{g(t) : t \in [0, T]\}$;

(3) There are continuous functions $A(t, x), C(t, x) > 0$ which are bounded on compact subsets of $[0, T] \times R$ and a constant $B \in [0, 1]$ such that

$$|f(t, x, r, q)| \leq A(t, x)r^2 + B|q| + C(t, x)$$

for all $(t, x) \in [0, T] \times [-M, M]$ and $r, q \in R$.

Then the problem (E 4.5.1) has a solution.

Proof. Take $r > \{M, M_1, M_2\}$, where M, M_1, M_2 are the same as in Lemma 4.5.5. Put $\Omega = B(0, r) = \{x(\cdot) : \|x(\cdot)\|_2 < r\}$. Then $Sx \neq \lambda Nx - \lambda g$ for all $\lambda \in (0, 1]$ and $x \in \partial\Omega$. Set $Qu = \frac{1}{T}\int_0^T u(t)dt$ for all $u \in Y$, then $Q : Y \to N(S)$ is a projection.

We define a bilinear form on $Y \times X$ by

$$[u, x] = \int_0^T u(t)x(t)dt \quad \text{for all } (u, x) \in Y \times X.$$

If $x(\cdot) \in \partial\Omega \cap N(S)$, then $\|x\|_2 = r > M$, so $x \equiv r$ or $-r$. Thus the assumption (2) implies that

$$QN(c) - Qg = \int_0^T [f(t, c, 0, 0) - g(t)]dt \neq 0$$

and

$$[QN(c) - Qg, c] = \int_0^T [f(t, c, 0, 0) - g(t)]cdt = 0,$$

where $c = r$ or $-r$. From Theorem 4.3.3, we know that the problem (E 4.5.1) has a solution. This completes the proof.

A special case of (E 4.5.1) is the following:

$$\begin{cases} x''(t) = f(t, x(t), x'(t)) - g(t), \quad t \in [0, T], \\ x(0) = x(T), \; x'(0) = x'(T). \end{cases} \tag{E 4.5.3}$$

In this case, the mapping given by $N(x)(t) = f(t, x(t), x'(t))$ for all $t \in [0, T]$ is compact, so $S - \lambda N$ is A-proper with respect to Π_S. From Theorem 4.5.6, the following immediately holds:

Corollary 4.5.7. Let $g(\cdot) \in Y$, $f(t, x, r) : [0, T] \times R^2 \to R$ be a continuous function and S, Π_S, N be the same as above. Suppose that the following conditions hold:

(1) There exist constants $M > 0$ and $c, d \in R$ with $c \le g_m \le g_M \le d$, and $x \le M$ implies that $d < f(t,x,0)$ for all $t \in [0,T]$, while $x \le -M$ implies that $f(t,x,0) < c$ for all $t \in [0,T]$, where $g_m = \min\{g(t) : t \in [0,T]\}$ and $g_M = \max\{g(t) : t \in [0,T]\}$;

(2) There are continuous functions $A(t,x), C(t,x) > 0$ which are bounded on compact subsets of $[0,T] \times R$ such that

$$|f(t,x,r)| \le A(t,x)r^2 + C(t,x) \text{ for all } (t,x) \in [0,T] \times [-M,M].$$

Then the problem (E 4.5.3) has a solution.

Next, we consider the wave equation:

$$\begin{cases} u_{tt}(t,x) - u_{xx}(t,x) - h(u(t,x)) = f(t,x), \\ t \in (0,2\pi),\ x \in (0,\pi), \\ u(t,0) = u(t,\pi) = 0, \quad t \in (0,2\pi), \\ u(0,x) = u(2\pi,x), \qquad x \in (0,\pi), \end{cases} \tag{E 4.5.4}$$

where $h : R \to R$ is a continuous function satisfying

$$|h(u)| \le \delta|u| + \gamma \tag{4.5.4}$$

and $f(\cdot) \in L^2((0,2\pi) \times (0,\pi))$, where $\delta > 0$ and $\gamma > 0$ are constants.

We say that $u \in L^2((0,2\pi) \times (0,\pi))$ is a weak solution of the problem (E 4.5.4) if

$$(u, v_{tt} - v_{xx}) - (h(u(t,x)), v) = (f(t,x), v)$$

for all $v \in C^2([0,2\pi] \times [0,\pi])$ with $v(t,0) = v(t,\pi) = 0$ for all $t \in [0,2\pi]$ and $v(2\pi,x) = v(0,x)$ for all $x \in [0,\pi]$.

Let $L : D(L) \subset L^2((0,2\pi)\times(0,\pi)) \to L^2((0,2\pi)\times(0,\pi))$ be the wave operator $Lu = u_{tt} - u_{xx}$. Then it is well known that L is self-adjoint, densely defined, and closed, and $Ker(L)$ is infinite dimensional with $Ker(L)^\perp = Im(L)$. Thus L is a Fredholm mapping of zero index type. Let $P : L^2((0,2\pi) \times (0,\pi)) \to Ker(L)$ be the projection. Then $(L + \lambda P)^{-1} : L^2((0,2\pi) \times (0,\pi)) \to D(L)$ is compact for all $\lambda > 0$.

For each $\eta > 0$, consider the following equation:

$$\begin{cases} u_{tt}(t,x) - u_{xx}(t,x) + \eta u(t,x) - h(u(t,x)) = f(t,x), \\ t \in (0,2\pi), x \in (0,\pi), \\ u(t,0) = u(t,\pi) = 0, \quad t \in (0,2\pi), \\ u(0,x) = u(2\pi,x), \qquad x \in (0,\pi), \end{cases} \tag{E 4.5.5}$$

where h, f is the same as in (E 4.5.4). Let u_η be the weak solution of (E 4.5.5) if it exists, and we set $S = \{u_\eta : \eta > 0\}$.

Theorem 4.5.8. One of the following conclusion holds:
(1) The problem (E 4.5.4) has a weak solution;
(2) S is unbounded in $L^2((0, 2\pi) \times (0, \pi)$.

Proof. Let $N : L^2((0, 2\pi) \times (0, \pi)) \to L^2((0, 2\pi) \times (0, \pi))$ be defined by

$$Nu(t, x) = h(u(t, x)) + f(t, x) \quad \text{for all } u(t, x) \in L^2((0, 2\pi) \times (0, \pi)).$$

By (4.5.4), N is a bounded and continuous mapping. We may assume that (2) does not hold, i.e., S is bounded, so there exists $r_0 > 0$ such that

$$\|u_\eta\|_{L^2} < r_0 \text{ for all } u_\eta \in S. \tag{4.5.5}$$

Let $\Omega = \{u(t, x) \in L^2((0, 2\pi) \times (0, \pi)) : \|u\|_{L^2} < r_0\}$. By (4.5.5), we know that $PNu \neq \eta Pu$ for all $u \in C^2([0, 2\pi] \times [0, \pi]) \cap \partial\Omega$ and $\eta > 0$. We may assume that $Lu \neq Nu$ for all $u \in C^2([0, 2\pi] \times [0, \pi]) \cap \partial\Omega$. By Theorem 4.4.16, we have $deg(L - N, \Omega, 0) = \{1\}$. Thus the problem ($E$ 4.5.4) has a weak solution. This completes the proof.

Remark. The results of Sections 4.1-4.3 in this chapter can be found in [239].

4.6 Exercises

1. Let X be a separable Banach space with a projection scheme $\Pi = \{X_n, P_n\}$, $\Omega \subset X$ be an open bounded subset with $x_0 \in \Omega$ and $T : \overline{\Omega} \to X$ be a P_1 compact mapping satisfying

$$Tx - x_0 \neq \lambda(x - x_0) \text{ for all } x \in \partial\Omega,\ \lambda > 1.$$

Show that T has a fixed point in $\overline{\Omega}$.

2. Let X be a separable Banach space with a projection scheme $\Pi = \{X_n, P_n\}$, $\Omega \subset X$ be an open bounded subset with $0 \in \Omega$ and $t\overline{\Omega} \subset \Omega$ for $\lambda \in (0, 1)$ and $T : \overline{\Omega} \to X$ be a P_1 compact mapping satisfying $T(\partial\Omega) \subseteq \overline{\Omega}$. Show that T has a fixed point in $\overline{\Omega}$.

3. Let X be a separable Banach space with a projection scheme $\Pi = \{X_n, P_n\}$, $\Omega \subset X$ be an open bounded subset with $0 \in \Omega$ and $T : \overline{\Omega} \to X$ be a P_1 compact mapping satisfying

$$\|x - Nx\|^2 \geq \|Nx\|^2 - \|x\|^2 \text{ for all } x \in \partial\Omega.$$

Show that T has a fixed point in $\overline{\Omega}$.

4. Let H be a real separable Hilbert space and $T : H \to H$ be a continuous mapping satisfying $(Tx - Ty, x - y) \geq c\|x - y\|^2$ for all $x, y \in H$ and some $c > 0$. Show that T is A-proper.

5. Let X be a separable Banach space with a projection scheme $\Pi = \{X_n, P_n\}$, $\Omega \subset X$ be an open bounded subset and $T : \overline{\Omega} \to X$ be a continuous A-proper mapping. Show that $T^{-1}y$ for all $y \in T(\overline{\Omega})$ is compact.

6. Let X be a separable Banach space with a projection scheme $\Pi = \{X_n, P_n\}$, $c = \sup_n \|P_n\| = c$, $D \subset X$ be closed and $T : D \to X$ be a mapping satisfying $\beta(T(B)) \leq k\beta(B)$ for all bounded subset B of D. Show that $\lambda I - T$ is A-proper for each $\lambda > kc$.

7. Let X be a separable Banach space with an operator scheme $\Pi = \{X_n, P_n\}$ and $T : X \to X$ be a continuous strongly accretive mapping, i.e.,
$$(x - y, Tx - Ty)_+ \geq c\|x - y\|^2 \quad \text{for all } x, y \in X, \ \text{ some } c > 0.$$
Show that T is A-proper.

8. Let X be a separable Banach space with an operator scheme $\Pi = \{X_n, P_n\}$ and $T : X \to X$ be a linear bounded operator. Show that T is A-proper and one to one if and only if $R(T) = X$ and $\|P_n Tx\| \geq c\|x\|$ on X_n for some $c > 0$ and all $n \geq n_0$.

9. Let $L : D(L) \subset X \to Y$ be a Fredholm mapping of zero index type and $\{Y_n, Q_n\}$ be a projection scheme for Y. Assume that X is reflexive. Show that, if $G \subset X$ is bounded closed, convex and $N : G \to Y$ is continuous compact, then $L + kN : G \cap D(L) \to Y$ is pseudo A-proper with respect to $\Gamma_{\lambda,L}$ for each $\lambda \neq 0$.

10. Let $L : D(L) \subset X \to Y$ be a Fredholm mapping of zero index, $\{Y_n, Q_n\}$ be a projection scheme for Y and $\Omega \subset$ be an open bounded subset. Let $N : \overline{\Omega} \to Y$ be a bounded mapping such that the following conditions are satisfied:

    (1) $L - \lambda N$ is A-proper with respect to Γ_L for each $\lambda \in (0, 1]$;

    (2) There are sets $E_1 \subset E_2 \subset R(L)$ such that $0 \in E_1$ and $\lambda E_2 \subset E_1$ for all $\lambda \in (0, 1)$ and, for all $x \in \partial(\Omega \cap D(L))$ such that $Nx \in R(L)$, one has $Lx \notin E_1$ and $Nx \in E_2$.

    If $Deg(L - QN, D(L) \cap \Omega, 0) \neq 0$, show that $Lx - Nx = 0$ has a solution, where Q is as in Section 4.1.

11. Let $L : D(L) \subset X \to Y$ be a Fredholm mapping of zero index, $\{Y_n, Q_n\}$ be a projection scheme for Y, $\Omega \subset$ be an open bounded subset. Let $N : \overline{\Omega} \to Y$ be a bounded mapping such that the following conditions are satisfied:

(1) $L - \lambda N$ is A-proper with respect to Γ_L for each $\lambda \in (0, 1]$;

(2) $Nx \notin R(L)$ for all $x \in L^{-1} \cap \partial\Omega$;

(3) $\|Lx - Nx\|^2 \geq \|Nx\|^2 - \|Lx\|^2$ for all $x \in D(L) \setminus (N(L) \cap \partial\Omega)$;

(4) $Deg(L - QN, D(L) \cap \Omega, 0) \neq 0$.

Show $Lx = Nx$ has a solution in $\Omega \cap D(L)$.

Chapter 5

COINCIDENCE DEGREE THEORY

In the 1970s, Mawhin systematically studied a class of mappings of the form $L + T$, where L is a Fredholm mapping of index zero and T is a nonlinear mapping, which he called a L- compact mapping. Based on the Lyapunov-Schmidt method, he was able to construct a degree theory for such mapping.

The goal of this chapter is to introduce Mawhin's degree theory for L-compact mappings. This chapter has four main sections.

We present some introductory material on Fredholm mappings and their relations with A-proper mappings in Section 5.1.

In Section 5.2, we define L-compact mappings (here L is a Fredholm mapping) and we introduce the coincidence degree. Various properties of this degree (see Theorem 5.2.2 and Lemma 5.2.6) are also discussed in this section.

In Section 5.3, various consequences of the degree theory in Section 5.2 are presented (see, in particular, Theorem 5.3.5).

An application to the periodic ordinary differential equations is presented in Section 5.4.

5.1 Fredholm Mappings

Definition 5.1.1. Let X and Y be normed spaces. A linear mapping $L : D(L) \subset X \to Y$ is called a Fredholm mapping if

(1) $Ker(L)$ has finite dimension;

(2) $Im(L)$ is closed and has finite codimension.

Proposition 5.1.2. Let X be a Banach space and $T : X \to X$ be a linear bounded mapping. Then $dim(Ker(T)) < \infty$ and $Im(T)$ is closed if and only if, for $x_n \in \overline{B(0,1)}$ such that $Tx_n \to y$, thus $(x_n)_{n=1}^{\infty}$ has a convergent subsequence.

Proof. For the "if" part, we know from our assumption that $\{x : Tx = 0, \|x\| \leq 1\}$ is compact and thus $Ker(T)$ is finite dimensional. We have $X = Ker(T) \oplus M$ for some closed subspace M of X. Obviously, we have $T(M) = Im(T)$. Since $T : M \to Im(T)$ is one to one, it follows that

$$\|Tx\| \geq c\|x\| \quad \text{for all } \ x \in M, \ \text{ some } \ c > 0$$

and, from this, we deduce that $T(M)$ is closed. Therefore, $Im(T)$ is closed.

For the "only if" part, assume that $x_n \in \overline{B(0,1)}$ such that $Tx_n \to y$. As before, $X = Ker(T) \oplus M$, so $x_n = z_n + m_n$ for some $x_n \in Ker(T)$ and $m_n \in M$. Thus $Tm_n \to y$. However, the restriction of T to M is continuous, one to one and onto $Im(T)$, so $m_n \to m \in M$. Recall that $dim(N(T))$ is finite, so $(x_n)_{n=1}^{\infty}$ has a convergent subsequence. This completes the proof.

Proposition 5.1.3. Let X be a Banach space, $T : X \to X$ be a linear bounded Fredholm operator and $K : X \to X$ be a linear continuous compact mapping. Then $T + K$ is a Fredholm mapping.

Proof. The proof is left to the reader as an exercise.

Recall that the codimension of $Im(L)$ is the dimension of $Coker(L) = Y/Im(L)$. If L is a Fredholm mapping, then its index is defined by

$$Ind(L) = dim(Ker(L)) - dim(Coker(L)).$$

Now, assume that L is a Fredholm mapping. Then there exist two linear continuous projections $P : X \to X$ and $Q : Y \to Y$ such that

$$Im(P) = Ker(L), \quad Ker(Q) = Im(L).$$

Also, we have

$$X = Ker(L) \oplus Ker(P), \quad Y = Im(L) \oplus Im(Q)$$

as the topological direct sums.

Obviously, the restriction of L_P of L to $D(L) \cap Ker(P)$ is one to one and onto $Im(L)$ and so its inverse $K_P : Im(L) \to D(L) \cap Ker(P)$ is defined. We denote by $K_{PQ} : Y \to D(L) \cap Ker(P)$ the generalized inverse of L defined by $K_{PQ} = K_P(I - Q)$.

Proposition 5.1.4. Let X, Y be separable Banach spaces and $L : D(L) \subset X \to Y$ be a densely defined Fredholm mapping with $Ind(L) = m \geq 0$. Then there exist a sequence of monotonically increasing finite dimensional subspaces $(X_n)_{n=1}^{\infty} \subset D(L)$ such that $\cup_{n=1}^{\infty} X_n$ is dense in X, and let $(P_n)_{n=1}^{\infty}$ be a sequence of linear continuous projections on X with $Im(P_n) = X_n$ for each $n \geq 1$ and $P_n x \to x$ for all $x \in X$ as $n \to \infty$ and $(Q_n)_{n=1}^{\infty}$ be a sequence of linear continuous projections on Y with $Im(Q_n) = Y_n$ for each $n \geq 1$ such that $dim(X_n) - dim(Y_n) = m$ for each $n \geq 1$.

Proof. Since L is Fredholm, there exist two linear continuous projections $P : X \to X$ and $Q : Y \to Y$ such that

$$Im(P) = Ker(L), \quad Ker(Q) = Im(L),$$

$$X = Ker(L) \oplus Ker(P), \quad Y = Im(L) \oplus Im(Q).$$

By the assumption $D(L)$ is dense in X, we may choose a a sequence $(X_n)_{n=1}^{\infty}$ of monotonically increasing finite dimensional subspaces of $D(L)$ such that $\cup_{n=1}^{\infty} X_n$ is dense in X and $Ker(L) \subset X_n$, a sequence $(P_n)_{n=1}^{\infty}$ of linear continuous projections on X with $Im(P_n) = X_n$ for each $n \geq 1$ and $P_n x \to x$ for each $x \in X$ as $n \to \infty$. Obviously, $PP_n = P$, $P_n(Ker(P)) \subset Ker(P)$ and $(I - P_n)(X) \subset Ker(P)$. Set $Q_n = Q + LP_n K_{PQ}$, then Q_n is continuous. It is easy to check that $Q_n^2 = Q_n$ for each $n \geq 1$. Thus, Q_n is a linear continuous projection for each $n \geq 1$. Finally, set $Y_n = Q_n(Y)$; then $Im(Q) \subset Q_n$, $Q_n Lx = LP_n x$ for all $x \in D(L)$ and $dim(X_n) - dim(Y_n) = m$ for each $n \geq 1$. This completes the proof.

In the sequel, we denote the approximation scheme constructed in Proposition 5.1.4 by $\Gamma_m = \{X_n, P_n; Y_n, Q_n\}$.

Definition 5.1.5. Let $G \subset X$ be a non-empty set, $G_n = G \cap X_n$ for $n = 1, 2, \cdots$. A mapping $T : G \to Y$ is said to be A-proper (respectively, pseudo A-proper) with respect to Γ_m if $T_n = Q_n T : G_n \to Y_n$ is continuous and, if $x_{n_k} \in G_{n_k}$ such that $(x_{n_k})_{k=1}^{\infty}$ is bounded and $Q_{n_k}(Tx_{n_k} - g) \to 0$ as $k \to \infty$ for some $g \in Y$, then there exists a subsequence $(x_{n_{k_l}})$ with $x_{n_{k_l}} \to x_0 \in G$ (resp., $x_0 \in G$ exists) such that $Tx = g$.

Proposition 5.1.6. Let X, Y be separable Banach spaces and $L : D(L) \subset X \to Y$ be a densely defined Fredholm mapping of index $m > 0$. Then L is A-proper with respect to Γ_m.

Proof. Let $x_{n_k} \in G_{n_k}$ be such that $(x_{n_k})_{k=1}^{\infty}$ is bounded and $Q_{n_k}(Tx_{n_k} - g) \to 0$ as $k \to \infty$ for some $g \in Y$. Notice that $Q_{n_k} = Q + LP_{n_k} K_{PQ}$, so we get

$$g_{n_k} = Q_{n_k}(Tx_{n_k} - g) = Lx_{n_k} - Qg - LP_{n_k} K_{PQ} g \to 0.$$

So $L(x_{n_k} - P_{n_k} K_{PQ} g) = g_{n_k} + Qg \to Qg$. However, $Im(L)$ is closed, so we have $Qg = 0$ and $L(x_{n_k} - P_{n_k} K_{PQ} g) \to 0$. From which we deduce that

$$K_P L(x_{n_k} - P_{n_k} K_{PQ} g) = (I - P)(x_{n_k} - P_{n_k}) K_{PQ} g) \to 0.$$

Now, the compactness of P implies that (x_{n_k}) has a subsequence $(x_{n_{k_l}})$ such that $x_{n_{k_l}} \to x_0$ and $x_0 - K_{PQ} g = Px_0 - PK_{PQ} g$. Since L is closed, we have $L(x_0 - K_{PQ} g) = 0$. Thus $Lx_0 = g$. This completes the proof.

More precisely, we have the following result between the A-proper mapping and the Fredholm mapping:

Proposition 5.1.7. Let X, Y be separable Banach spaces and $L : X \to Y$ be a linear bounded mapping. Then L is a Fredholm mapping of index $m \geq 0$ if and only if L is A-proper with respect to some projectional scheme.

Proposition 5.1.8. Let X, Y be Banach spaces, $L : D(L) \subset X \to Y$ be a Fredholm mapping of index $m > 0$, $J' : Im(Q \to Ker(L)$ be a monomorphism and $N : D(N) \subset X \to Y$ be a mapping. Then $Lx - Nx = y$ if and only if

$$x - Px - J'QNx - K_{PQ}Nx = K_{PQ}y + J'Qy.$$

Proof. If $Lx - Nx = y$, then we have

$$L(I - P)x - QNx - (I - Q)Nx = Qy + (I - Q)y,$$

so $-QNx = Qy$ and $L(I-P)x-(I-Q)Nx = (I-Q)y$, i.e., $-J'QNx = JQy$ and $x - Px - K_{PQ}Nx = K_{PQ}y$. Thus we have

$$x - Px - J'QNx - K_{PQ}Nx = K_{PQ}y + J'Qy.$$

On the other hand, if $x - Px - J'QNx - K_{PQ}Nx = K_{PQ}y + J'Qy$, then, since $J'Nx \in Ker(L) = Im(P)$ and $K_{PQ}Nx \in D(L) \cap Ker(P)$, we have $x - Px - K_{PQ}Nx = K_{PQ}y$ and $-J'QNx = J'Qy$ and so $Lx - (I - Q)Nx = (I - Q)y$ and $-QNx = Qy$. Thus $Lx - Nx = y$. This completes the proof.

Proposition 5.1.9. Let X, Y be separable Banach spaces, $L : D(L) \subset X \to Y$ be a densely defined Fredholm mapping of index $m > 0$ and Γ_m the same as Proposition 5.1.4. Let $\Omega \subset X$ be an open bounded subset and $N : \overline{\Omega} \to Y$ be a bounded mapping. If $I - P - J'QN - K_{pQ}N$ is A-proper with respect to $\Gamma = \{X_n, P_n\}$, then $L - N$ is A-proper with respect to Γ_m.

Proof. For any $x_{n_k} \in \overline{\Omega} \cap D(L) \cap X_{n_k}$ such that $Q_{n_k}(Lx_{n_k} - Nx_{n_k} - y \to 0$ as $k \to \infty$ for some $y \in Y$, we recall that $Q_{n_k} = Q + LP_{n_k}K_{pQ}$, so we have

$$\begin{aligned}
&Q_{n_k}(Lx_{n_k} - Nx_{n_k} - y) \\
&= L(I - P_{n_k})x_{n_k} - Q(Nx_{n_k} + y) - LP_{n_k}(I - Q)(Nx_{n_k} + y) \\
&\to 0
\end{aligned}$$

as $k \to \infty$. Therefore, we have

$$y_k = L(I - P_{n_k})x_{n_k} - LP_{n_k}(I - Q)(Nx_{n_k} + y) \to 0,$$

$$z_k = QNx_{n_k} + Qy \to 0$$

and thus

$$h_k = K_P y_k = (I - P)x_{n_k} - P_k K_{PQ}(Nx_{n_k} + y) \to 0,$$

$$w_k = J'z_k = J'QNx_{n_k} + J'Qy \to 0.$$

From which we deduce that

$$x_{n_k} - Px_{n_k} - P_{n_k}K_pNx_{n_k} \to K_{PQ}y,$$

$$J'QNx_{n_k} \to -J'Qy,$$

which immediately implies that

$$P_{n_k}(I - P - J'QN - K_{pQ}N)x_{n_k} \to K_{pQ}y + J'Qy.$$

So $(x_{n_k})_{k=1}^{\infty}$ has a convergence subsequence $(x_{n_{k_l}})$ with $x_{n_{k_l}} \to x_0$ and $x_0 - Px_0 - J'QNx_0 - K_{pQ}Nx_0 = K_{pQ}y + J'Qy$. By Proposition 5.1.8, we have $Lx_0 - Nx_0 = y$. Thus $L - N$ is A-proper with respect to Γ_m. This completes the proof.

Assume now that L is a Fredholm mapping of index zero. Then, for any isomorphism $J : Im(Q) \to Ker(L)$, the mapping $JQ + K_{PQ}$ is an isomorphism from Y onto $D(L)$ and

$$(JQ + K_{PQ})^{-1}x = (L + J^{-1}P)x \quad \text{for all } x \in D(L).$$

In fact, if $y \in Y$, we have

$$(JQ + K_{PQ})y = x \;\Leftrightarrow\; JQy = Px,$$

$$K_{PQ}y = (I - P)x \;\Leftrightarrow\; Qy = J^{-1}Px,$$

$$L_PK_{PQ}y = L(I - P)x \;\Leftrightarrow\; Qy = J^{-1}Px,$$

$$(I - Q)y = Lx \;\Leftrightarrow\; y = (J^{-1}P + L)x.$$

Example 5.1.10. Let X be a real Banach space and $T : X \to X$ be a linear continuous compact mapping. Then, by (5) of Theorem 2.1.15, we know that $dim(ker(I + T)) = dim(codim(I + T)) < +\infty$, so $I + T$ is a Fredholm mapping of index zero.

Example 5.1.11. Let $f : [0, T] \to R$ be in L^1 and consider the following problem:

$$\begin{cases} x'(t) = f(t), & t \in (0, T), \\ x(0) = x(T). \end{cases} \qquad (E\ 5.1.1)$$

We set $X = C([0, T], R)$, the space of all continuous function from $[0, T]$ to R, $Y = L^1([0, T], R) \times R$, $y_0 = (f(\cdot), 0)$ and define a mapping $L : X \to Y$ by

$$Lx(\cdot) = (x'(\cdot), x(0) - x(T)) \quad \text{for all } x(\cdot) \in dom(L),$$

where $dom(L) = \{x(\cdot) \in C([0, T], R) : x'(\cdot) \in L^1([0, T], R)\}$.

One can easily see that $(E\ 5.1.1)$ is equivalent to the following equation:

$$Lx(\cdot) = y_0.$$

It is easy to see that

$$ker(L) = \{x(\cdot) \in C([0,T], R) : x(t) = c,\, t \in [0,T],\, c \in R\},$$

$$Im(L) = (f(\cdot), -\int_0^T f(s)ds) \quad \text{for all } f(\cdot) \in L^1([0,T], R).$$

Obviously, $Im(L)$ is closed and $dim(Ker(L)) = dim(Coker(L)) = 1$. Thus L is a Fredholm mapping of index zero.

5.2 Coincidence Degree for L-Compact Mappings

In this section, we define coincidence degree for L-compact mappings and give some properties of coincidence degree.

Definition 5.2.1. Let $L : D(L) \subset X \to Y$ be a Fredholm mapping, E be a metric space and $T : E \to Y$ be a mapping. We say that T is *L-compact* on E if $QT : E \to Y$, $K_{PQ}T : E \to X$ are continuous and $QT(E)$, $K_{PQ}T(E)$ are compact, where all notations are the same as in Section 5.1.

Now, we are ready to introduce the conincidence degree:

Coincidence degree. Let X, Y be real normed spaces, $L : D(L) \subset X \to Y$ be a Fredholm mapping of index zero and Ω be an open bounded subset of X. Suppose that $F = L + T : D(L) \cap \overline{\Omega} \to Y$ is a mapping and $T : \overline{\Omega} \to Y$ is L-compact on $\overline{\Omega}$. Suppose also that $0 \notin F(D(L) \cap \partial\Omega)$. Let $J : Im(Q) \to Ker(L)$ be an isomorphism. Put $H^J_{PQ} = JQ + K_{PQ}$. It is easy to check that

$$H^J_{PQ}F = K_{PQ}L + H^J_{PQ}T = I - P + (JQ + K_{PQ})T.$$

Consequently, $0 \notin H^J_{PQ}F(D(L) \cap \partial\Omega)$ (if $0 \in H^J_{PQ}F(D(L) \cap \partial\Omega$, then $0 = K_{PQ}(Lx + Tx) + JQTx$ for some $x \in D(L) \cap \partial\Omega$, so $QTx = 0$ and $(I - Q)(Lx + Tx) = 0$. Thus $Lx + Tx = 0$, which is a contradiction). By the L-compactness of T, the Leray Schauder degree $deg(I - P + (JQ + K_{PQ})T, \Omega, 0)$ is well defined.

Now, we define a degree by

$$D_J(L + T, \Omega, 0) = deg(I - P + (JQ + K_{PQ})T, \Omega, 0),$$

which is called the coincidence degree of L and $-T$ on $\Omega \cap D(L)$. One can easily prove that this definition does not depend on the choice of P, Q. It is known that $D_J(L + T, \Omega, 0)$ is a constant for some J depending on orientations on $Ker(L)$ and $Coker(L)$ (see [203]), so the coincidence degree in [203] is defined only for those $J's$. The definition given here depends on the J.

Remark. (1) If $dim(X) = dim(Y) < +\infty$ and we take $L = 0$, then any continuous mapping T on $\overline{\Omega}$ is L-compact. If we take $P = I$ and $Q = I$, then it follows that $K_{PQ} = 0$, so $H^J_{PQ}F = JT$ and thus we have

$$D_J(T, \Omega, 0) = deg(JT, \Omega, 0) = sign(detJ)deg(T, \Omega, 0).$$

Therefore, if we only take those J such that $detJ > 0$, then we have $D_J(T, \Omega, 0) = deg(T, \Omega, 0)$, which is the Brouwer degree.

(2) If $X = Y$ and we take $L = I$, then any continuous compact mapping T on $\overline{\Omega}$ is L-compact. If we take $P = Q = 0$, then $K_{PQ} = I$, $J = 0 : \{0\} \to \{0\}$ and $H^J_{PQ}F = I + T$. Thus $D_J(I + T, \Omega, 0) = deg(I + T, \Omega, 0)$, which is the Leray Schauder degree.

Theorem 5.2.2. The coincidence degree of L and $-T$ on Ω has the following properties:

(1) If Ω_1 and Ω_2 are disjoint open subsets of Ω such that $0 \notin F(D(L) \cap \Omega \setminus (\Omega_1 \cup \Omega_2)$, then

$$D_J(L + T, \Omega, 0) = D_J(L + T, \Omega_1) + D_J(L + T, \Omega_2, 0);$$

(2) If $H(t, x) : [0, 1] \times \overline{\Omega} \to Y$ is L-compact on $[0, 1] \times \overline{\Omega}$ and $0 \neq Lx + H(t, x)$ for all $(t, x) \in [0, 1] \times \partial\Omega$, then $D_J(L + H(t, \cdot), \Omega, 0)$ does not depend on $t \in [0, 1]$;

(3) If $D_J(L + T, \Omega, 0) \neq 0$, then $0 \in (L + T)(D(L) \cap \Omega)$.

Corollary 5.2.3. If T_1, T_2 are L-compact mappings on $\overline{\Omega}$ and $T_1x = T_2x$ for all $x \in D(L) \cap \partial\Omega$, then

$$D_J(L + T_1, \Omega, 0) = D_J(L + T_2, \Omega, 0).$$

Proof. We define $H(t, x) : [0, 1] \times \overline{\Omega} \to Y$ by

$$H(t, x) = tT_1x + (1 - t)T_2x \quad \text{for all } (t, x) \in [0, 1] \times \overline{\Omega}.$$

Then H is L-compact. Therefore, it follows from (3) of Theorem 5.2.2 that $D_J(L + T_1, \Omega, 0) = D_J(L + T_2, \Omega, 0)$.

Proposition 5.2.4. Let X, Y be real normed spaces, $L : D(L) \subset X \to Y$ be a Fredholm mapping of index zero, Y_0 be a finite dimensional subspace of Y satisfying $Y = Im(L) \oplus Y_0$ algebraically and Ω be an open bounded subset of X. If T is L-compact on $\overline{\Omega} \cap D(L)$, and $T(\overline{\Omega}) \subset Y_0$, then

$$D_J(L + T, \Omega, 0) = signdet(J)deg(T, \Omega \cap Ker(L), 0),$$

where $deg(\cdot, \cdot, \cdot)$ is the Brouwer degree.

Proof. Since L is a Fredholm mapping of index zero, we get $Y = Im(L) \oplus Y_0$ topologically. Take $Q : Y \to Y$ with $Im(Q) = Y_0$, then $QTx = Tx$ for $x \in \overline{\Omega}$ and

$$H^J_{PQ}F = (JQ + K_{PQ})(L + T) = I - P + JT.$$

Note that $(P - JT)(\overline{\Omega}) \subset Ker(L)$ and $I = P$ on $Ker(L)$; thus, by Theorem 2.2.9, we have

$$D_J(L + T, \Omega, 0) = deg(I - P + JT, \Omega, 0) = deg(I - P + JT, \Omega \cap Ker(L), 0).$$

But $I = P$ on $Ker(L)$ and thus we get

$$\begin{aligned} D_J(L + T, \Omega, 0) &= deg(JT, \Omega \cap Ker(L), 0) \\ &= signdet(J)deg(T, \Omega \cap Ker(L), 0). \end{aligned}$$

This completes the proof.

Lemma 5.2.5. Let $A : Im(Q) \to D(L)$ be a linear mapping such that $PA : Im(Q) \to Ker(L)$ is an isomorphism. Then $H^A_{PQ} = AQ + K_{PQ}$ is an algebraic isomorphism from Y onto $D(L)$ and

$$(H^A_{PQ})^{-1} = L - LA(PA)^{-1}P + (PA)^{-1}P.$$

Moreover, if T is L-compact on $\overline{\Omega}$, then

$$H^A_{PQ}(L + T) = I - P + H^A_{pQ}T$$

with $H^A_{PQ}T : \overline{\Omega} \to X$ continuous and compact.

Proof. For any $x \in D(L)$, $H^A_{PQ}z = x$ if and only if $PAQz = Px$ and $(I - P)AQz + K_{PQ}z = (I - P)x$, i.e.,

$$Qz = (PA)^{-1}Ox, LAQz + (I - Q)z = Lx,$$

$$z = Lx - LA(PA)^{-1}Px + (PA)^{-1}Px.$$

Finally, we have

$$\begin{aligned} H^A_{PQ}(L + T) &= (AQ + K_{PQ})(L + T) \\ &= K_{PQ}L + (AQ + K_{PQ})T \\ &= (I - P) + H^A_{PQ}T. \end{aligned}$$

By the assumption, QT and $K_{PQ}T$ are continuous and compact on $\overline{\Omega}$, so $H^A_{PQ}T$ is continuous and compact on $\overline{\Omega}$. This completes the proof.

Lemma 5.2.6. Let $A, B : Im(Q) \to D(L)$ be linear mappings such that PA, $PQ : Im(Q) \to Ker(L)$ are isomorphisms and T is L-compact. Then

$$\begin{aligned} &deg(I - P + H^B_{PQ}T, \Omega, 0) \\ &= deg(I - (A - B)(PA)^{-1}P, B(0, r), 0)deg(I - P + K^A_{PQ}T, \Omega, 0) \end{aligned}$$

for any $r > 0$.

Proof. First, we have

$$\begin{aligned}&(I-(A-B)(PA)^{-1}P)(I-P+K_{PQ}^{A}T)\\ &= I-P+AQ+K_{pQ})T-(A-B)(PA)^{-1}PAQT\\ &= I-P+H_{PQ}^{B}T.\end{aligned}$$

Therefore, it follows that

$$deg(I-P+H_{PQ}^{B}T,\Omega,0) = deg(I-(A-B)(PA)^{-1}P)(I-P+K_{PQ}^{A}T),\Omega,0).$$

Moreover, if $x-(A-B)(PA)^{-1}Px = 0$, then $PB(PA)^{-1}Px = 0$, so $Px = 0$. Thus $x = 0$ and so $I-(A-B)(PA)^{-1}P$ is a homeomorphism of X onto itself. By the product formula of Theorem 2.2.8, we know that

$$\begin{aligned}&deg(I-(A-B)(PA)^{-1}P)(I-P+K_{PQ}^{A}T),\Omega,0)\\ &= \Sigma_i deg(I-(A-B)(PA)^{-1}P,(U_i,0)deg(I-P+K_{PQ}^{A}T,\Omega,U_i),\end{aligned}$$

where U_i are connected components of $X \setminus (I-P+K_{pQ}^{A}T)(\partial\Omega)$.

There are now two cases:

Case (1) If $0 \in U_i$, then $0 \notin U_k$ for $k \neq i$. Thus $deg(I-(A-B)(PA)^{-1}P,U_k,0) = 0$, so we have

$$\begin{aligned}&deg(I-(A-B)(PA)^{-1}P)(I-P+K_{PQ}^{A}T),\Omega,0)\\ &= deg(I-(A-B)(PA)^{-1}P,U_i,0)deg(I-P+K_{PQ}^{A}T,\Omega,0).\end{aligned}$$

By the excision property of Leray Schauder degree, the conclusion is true.

Case (2) If $0 \notin U_i$ for all $i \geq 1$, then $deg(I-(A-B)(PA)^{-1}P,U_i,0) = 0$ for all $i \geq 1$ and thus the conclusion holds. This completes the proof.

From Lemma 5.2.6, the following holds immediately:

Corollary 5.2.7. Let T be L-compact on $\overline{\Omega} \cap D(L)$ and A be the same as in Lemma 5.2.6. Then

$$\begin{aligned}&D_J(L+T,\Omega,0)\\ &= deg(I-(A-J)(PA)^{-1}P,B(0,r))deg(I-P+H_{PQ}^{A}T,\Omega,0)\end{aligned}$$

for all $r > 0$.

Proposition 5.2.8. If $T: X \to Y$ is linear and L-completely continuous, $Ker(L+T) = \{0\}$ and $\Omega \subset X$ is a nonempty open bounded subset such that $0 \notin \partial\Omega$, then

$$|D_J(L+T,\Omega,0)| = \begin{cases}0, & 0 \notin \Omega,\\ 1, & 0 \in \Omega.\end{cases}$$

Proof. By definition, we have

$$D_J(L+T,\Omega,0) = deg(I-P+K_{PQ}T,\Omega,0).$$

Thus the conclusion follows from the assumption and Theorem 2.2.4.

Theorem 5.2.9. If $\Omega \subset X$ is open bounded with $0 \in \Omega$, Ω is symmetric with respect to 0 and T is L-compact on $\overline{\Omega} \cap D(L)$ such that $T(-x) = -Tx$ for all $x \in \partial\Omega \cap D(L)$, then $|D_J(L+T,\Omega,0)|$ is an odd number.

Proof. Since $D_J(L+T,\Omega,0) = deg(I-P+K_{PQ}T,\Omega,0)$, by the definition of the Leray Schauder degree and Borsuk's Theorem (Theorem 1.2.11), we know that the conclusion is true.

In the following, let $L : D(L) \subset X \to Y$ be a Fredholm mapping of index zero, and $L = L_1 + L_2$, where L_1, L_2 satisfying the following conditions:

(1) $L_1 : D(L) \to Y$ is a Fredholm mapping of index zero;

(2) $L_2 : X \to Y$ is linear and L_1-completely continuous on X.

Now, assume that $\Omega \subset X$ is a nonempty open bounded subset of X, $T : \overline{\Omega} \to Y$ is L-compact on $\overline{\Omega}$ and T is also L_1-compact on $\overline{\Omega}$. Set $T_1 = L_2 + T$. Then T_1 is L_1-compact on $D(L) \cap \overline{\Omega}$ and $L + T = L_1 + T_1$.

Let P, Q, J be the linear mapping associated with L as in Section 5.1 and P_1, Q_1, J_1 be the corresponding ones for L_1. Put $H = JQ + K_{PQ}$ and $H_1 = J_1Q_1 + K_{P_1Q_1}$. Then $H, H_1 : Y \to D(L)$ are algebraic isomorphisms, $H^{-1} = L + J^{-1}P$ and $H_1^{-1} = L_1 + J_1^{-1}P_1$. We set $K_1 = L_2 + J^{-1}P$. It is easy to check that $J^{-1}P$ is L_1-completely continuous and hence K_1 is also L_1-completely continuous.

We have seen that

$$H(L+T) = I - P + HT, \; H_1(L+T) = H_1(L_1+T_1) = I - P_1 + H_1T_1,$$

$$(I - P + HT)(D(L) \cap \overline{\Omega}) = H(L+T)(D(L) \cap \overline{\Omega}) \subset D(L),$$

so we have

$$I - P_1 + H_1T_1 = H_1H^{-1}(I - P + HT). \tag{5.2.1}$$

On the other hand, $H_1H^{-1} = H_1(L_1 + K_1) = (I - P_1 + H_1K_1)$. Therefore, we have come to the following conclusion:

Proposition 5.2.10. $I - P_1 + H_1T_1 = (I - P_1 + H_1K_1)(I - P + HT)$ and $I - P_1 + H_1K_1$ is a linear homeomorphism on X.

Proof. We have seen that

$$I - P_1 + H_1T_1 = (I - P_1 + H_1K_1)(I - P + HT).$$

Note that $Ker(I - P_1 + H_1T_1 = Ker(H_1H^{-1}) = \{0\}$ and P_1, H_1T_1 are compact linear mappings, so $I - P_1 + H_1K_1$ is a linear homeomorphism on X.

Corollary 5.2.11. Under the above assumptions, we have

$$|D_{J_1}(L_1 + T_1, \Omega, 0)| = |D_J(L + T, \Omega, 0)|.$$

Proof. By the definition of coincidence degree and Proposition 5.2.10, we have

$$\begin{aligned} & D_{J_1}(L_1 + T_1, \Omega, 0) \\ &= deg((I - P_1 + H_1K_1)(I - P + HT), \Omega, 0) \\ &= \Sigma_i deg(I - P_1 + H_1K_1, U_i, 0) deg(I - P + HT, \Omega, U_i), \end{aligned}$$

where U_i are connected components of $X \setminus (I - P + HT)(\partial\Omega)$.

We have the following two cases:

Case (1) If $0 \in U_i$, then $0 \notin U_k$ for $k \neq i$, so we have

$$D_{J_1}(L_1 + T_1, \Omega, 0) = deg(I - P_1 + H_1K_1, U_i, 0) deg(I - P + HT, \Omega, 0).$$

Since $I - P_1 + H_1K_1$ is a homeomorphism on X, by Theorem 2.2.4, we know that $|D_{J_1}(L_1 + T_1, \Omega, 0)| = |D_J(L + T, \Omega, 0)|$.

Case (2) If $0 \notin U_i$ for all $i \geq 1$, then $0 \notin (I - P + HT)(\Omega)$, so we have

$$|D_{J_1}(L_1 + T_1, \Omega, 0)| = |D_J(L + T, \Omega, 0)| = 0.$$

This completes the proof.

Let X, Y be real normed spaces, $L : D(L) \subset X \to Y$ be a Fredholm mapping of index zero, $\Omega \subset X$ be an open bounded subset and $T : \overline{\Omega} \to Y$ be L-compact. Assume that a is an isolated zero of $L - T$, then we define

$$i_J(L - T, a) = \lim_{r \to 0} D_J(L - T, B(a, r) \cap D(L), 0),$$

which is called the coincidence index of L and T at a. One may easily see that this definition is well defined by using the excision property of coincidence degree.

The following result follows immediately from the definition:

Proposition 5.2.12. Let $\Omega \subset X$ be an open bounded subset and $T : \overline{\Omega} \to Y$ be L-compact. If $(L - T)^{-1}(0) = \{a_1, a_2, \cdots, a_k\} \subset \Omega$, then

$$D_J(L - T, \Omega \cap D(L), 0) = \Sigma_{i=1}^k i_J(L - T, a_i).$$

Proposition 5.2.13. Let $\Omega \subset X$ be an open bounded subset with $0 \in \Omega$ and $T : \overline{\Omega} \to Y$ be L-compact. Suppose that $A : X \to Y$ is a linear mapping such that A is L-compact on any bounded subset of X. Assume that

$$\lim_{\|x\|\to 0} \frac{\|QTx\| + \|K_{PQ}Tx\|}{\|x\|} = 0, \tag{5.2.2}$$

and $Ker(L - A) = \{0\}$. Then 0 is an isolated zero of $L - A - T$ and

$$i_J(L - A - T, 0) = i_J(L - A, 0).$$

Proof. By assumption, A is L-compact on any bounded suubset, so $I - P - JQA - K_{PQ}A$ is a linear completely continuous perturbation of the identity. Also, $Ker(I - P - JQA - K_{PQ}A) = \{0\}$ and thus there exists $c > 0$ such that

$$\|(I - P - JQA - K_{PQ}A)x\| \geq c\|x\|, x \in X.$$

From the assumption (5.2.2), we know that there exists $r > 0$ such that

$$B(0, r) \subset \Omega, \quad \|QTx\| + \|K_{PQ}Tx\| \leq 2^{-1}c\|x\| \text{ for all } x \in \overline{B(0, r)},$$

which implies that

$$\|(I - P - JQA - K_{PQ}A - tJQT - tK_{PQ}T)x\| \geq 2^{-1}c\|x\|$$

for all $x \in \overline{B(0, r)}$ and $t \in [0, 1]$. Thus we have

$$D_J(L - A - T, B(0, r) \cap D(L), 0) = D_J(L - A, B(0, r) \cap D(L), 0),$$

i.e., $i_J(L - A - T, 0) = i_J(L - A, 0)$. This completes the proof.

5.3 Existence Theorems for Operator Equations

Let X, Y be real normed spaces, $L : D(L) \subseteq X \to Y$ be a linear Fredholm mapping of index zero and $\Omega \subset X$ be an open bounded subset with $D(L) \cap \overline{\Omega} \neq \emptyset$.

Theorem 5.3.1. Let $0 \in \Omega$, and Ω symmetric with respect to 0 and $T : \overline{\Omega} \to Y$ be L-compact. If $Lx - Tx \neq t(-Lx - T(-x))$ for all $(t, x) \in (0, 1] \times D(L) \cap \partial\Omega$, then $Lx - Tx = 0$ has a solution in $D(L) \cap \overline{\Omega}$.

Proof. Let a mapping $H(t, x) : [0, 1] \times \overline{\Omega} \to Y$ be defined by

$$H(t, x) = \frac{1 + t}{2}Tx - \frac{1 - t}{2}T(-x) \quad \text{for all } (t, x) \in [0, 1] \times \overline{\Omega}.$$

Then $H(t,\cdot)$ is a homotopy of L-compact mappings. If $Lx - H(t,x) = 0$ for some $(t,x) \in [0,1) \times D(L) \cap \partial\Omega$, then

$$Lx - Tx = \frac{1-t}{1+t}(-Lx - T(-x)),$$

which is a contradiction. We may also assume that $Lx - Tx \neq 0$ for $x \in \partial\Omega$. Otherwise, the conclusion is true. By Theorem 5.2.9, we have $D_J(L - T, \Omega, 0) \neq 0$, thus $Lx - Nx = 0$ has a solution in $D(L) \cap \overline{\Omega}$. This completes the proof.

Theorem 5.3.2. Let $T_1, T_2 : \overline{\Omega} \to Y$ be L-compact. If the following conditions are satisfied:

(1) $Lx - tT_1x + (1-t)T_2x \neq 0$ for all $(t,x) \in (0,1) \times D(L) \cap \partial\Omega$;

(2) $D_J(L + T_2, \Omega, 0) \neq 0$;

then $Lx - T_1x$ has a solution in $D(L) \cap \overline{\Omega}$.

Proof. We may also assume that $Lx - T_1x \neq 0$ for all $x \in \partial\Omega$. Otherwise, the conclusion is true. Let $H(t,x) : [0,1] \times \overline{\Omega} \to Y$ be defined by

$$H(t,x) = tT_1x - (1-t)T_2x \quad \text{for all } (t,x) \in (t,x) \in [0,1] \times \overline{\Omega}.$$

By assumption, we have $Lx - H(t,x) \neq 0$ for all $(t,x) \in [0,1] \times D(L) \cap \partial\Omega$. Thus we have

$$D_J(L - T_1, \Omega, 0) = D_J(L - T_2, \Omega, 0) \neq 0$$

and so $Lx - T_1x$ has a solution in $D(L) \cap \overline{\Omega}$. This completes the proof.

Theorem 5.3.3. Let $T_1, T_2 : \overline{\Omega} \to Y$ be L-compact. If $L + T_2$ is one to one on $\overline{\Omega}$ and

$$Lx - t_1Tx + (1-t)(T_2x - p) \quad \text{for all } (t,x) \in (0,1) \times D(L) \cap \partial\Omega,$$

where $p \in (L + T_2)(D(L) \cap \Omega)$, then $Lx - T_1x$ has a solution in $D(L) \cap \overline{\Omega}$.

Proof. Since $L + T_2$ is one to one and $p \in (L + T_2)(D(L) \cap \Omega)$, we have $|D_J(L + T_2 - p, \Omega, 0)| = 1$. Thus, the conclusion follows from Theorem 5.3.2.

Theorem 5.3.4. Let $A : X \to Y$ be a linear continuous L-compact mapping with $Ker(L - A) = \{0\}$ and $T : \overline{\Omega} \to Y$ be L-compact. Suppose that the following conditions hold:

(1) $0 \in \Omega$ and $\lambda\partial\Omega \subset \Omega$ for all $\lambda \in (0,1)$;

(2) $(T - A)(D(L) \cap \partial\Omega) \subset (L - A)(D(L) \cap \Omega$.

Then $Lx = Tx$ has a solution in $D(L) \cap \overline{\Omega}$.

Proof. Put $H(t,x) = (1-t)Ax + tTx$ for all $(t,x) \in [0,1] \times \overline{\Omega}$. We claim that

$$Lx \neq H(t,x) \quad \text{for all } (t,x) \in (0,1) \times D(L) \cap \partial\Omega.$$

If this is not true, then there exist $(t,x) \in (0,1) \times D(L) \cap \partial\Omega$ such that $Lx \neq H(t,x)$. Then we have

$$\begin{aligned}(L-A)x &= \lambda(T-A)x \\ &\in \lambda(T-A)(D(L) \cap \partial\Omega \\ &\subset \lambda(T-A)(D(L) \cap \overline{\Omega} \\ &= (L-A)(D(L) \cap \lambda\overline{\Omega} \\ &\subset (L-A)(D(L) \cap \Omega),\end{aligned}$$

which is impossible because $L - A$ is one to one.

By assumption and Proposition 5.2.8, $|D_J(L-A, \Omega, 0)| = 1$. If $Lx - Tx = 0$ for some $x \in D(L) \cap \partial\Omega$, then the conclusion is true. Otherwise, we have

$$D_J(I-T, \Omega, 0) = D_J(L-A, \Omega, 0) \neq 0,$$

thus $Lx - Tx = 0$ has a solution in $D(L) \cap \overline{\Omega}$. This completes the proof.

Theorem 5.3.5. Let $T_1, T_2 : \overline{\Omega} \to Y$ be L-compact. Let $Z \subset Y$ be a subspace with $Y = Im(L) \oplus Z$ algebraically and $T_2(\overline{\Omega}) \subset Z$. Suppose that the following conditions hold:

(1) $Lx - (1-t)T_2 - tT_1 \neq 0$ for all $(t,x) \in (0,1) \times D(L) \cap \partial\Omega$;

(2) $T_2 x \neq 0$ for all $x \in Ker(L) \cap \partial\Omega$;

(3) $deg(T_{Ker(L)}, \Omega \cap Ker(L), 0) \neq 0$, where $T_{Ker(L)}$ is the restriction of T_2 to $Ker(L) \cap \overline{\Omega}$.

Then $Lx = T_1 x$ has a solution in $D(L) \cap \overline{\Omega}$.

Proof. Put $H = L - T_2 x$ and let $Q : Y \to Y$ be the projection such that $Im(Q) = Z$ and $Ker(Q) = Im(L)$. Then $QT_2 = T_2$ and $Hx = 0$ if and only if $QHx = 0, (I-Q)Hx = 0$, i.e., $T_2 x = 0$ and $Lx = 0$. Therefore, by the assumption (2) and Proposition 5.2.4, we have

$$|D_J(L - T_2, \Omega, 0)| = |deg(T_{Ker(L)}, \Omega \cap Ker(L), 0)| \neq 0.$$

Thus, it follows from Theorem 5.3.2 that $Lx = T_1 x$ has a solution in $D(L) \cap \overline{\Omega}$. This completes the proof.

Corollary 5.3.6. Let $T : \overline{\Omega} \to Y$ be L-compact. Suppose that the following conditions hold:

(1) $Lx - tT \neq 0$ for all $(t, x) \in (0, 1) \times (D(L) \setminus Ker(L)) \cap \partial\Omega$;

(2) $Tx \notin Im(L) = 0$ for all $x \in Ker(L) \cap \partial\Omega$;

(3) $deg(QT_{Ker(L)}, \Omega \cap Ker(L), 0) \neq 0$, where $Q : Y \to Y$ is the projection such that $Ker(Q) = Im(L)$.

Then $Lx = Tx$ has a solution in $D(L) \cap \overline{\Omega}$.

Proof. Put $Z = Im(Q)$ and $T_2 = QT$ in Theorem 5.3.5. By the assumption (2), we know that

$$QTx \neq 0 \quad \text{for all } x \in Ker(L) \cap \partial\Omega.$$

Now, if $Lx - (1-t)QTx - tTx = 0$ for some $(t, x) \in (0, 1) \cap D(L) \cap \partial\Omega$, then we have

$$QNx = 0, Lx - tTx = 0.$$

It easily follows that $x \in D(L) \setminus Ker(L)) \cap \partial\Omega$, which contradicts the assumption (1). Thus the conditions of Theorem 5.3.5 are satisfied and, consequently, $Lx = Tx$ has a solution in $D(L) \cap \overline{\Omega}$.

5.4 Applications to ODEs

In this section, we give some applications of the results to differential equations.

Let $f(t, x, y) : [0, \pi] \times R^n \times R^n \to R^n$ be a function satisfying the *Caradéodory condition*, i.e.,

(1) For almost all $t \in [0, \pi]$, $f(t, x, y)$ is continuous in (x, y);

(2) For all $(x, y) \in R^n \times R^n$, $f(t, x, y)$ is measurable in t;

(3) For all $r > 0$, there exists $g_r(\cdot) \in L^1([0, \pi], [0, +\infty))$ such that, For almost all $t \in [0, \pi]$,

$$|f(t, x, y)| \leq g_r(t) \quad \text{for all } x, y \in R^n, \ |x| \leq r, \ |x| \leq r.$$

Consider the Picard boundary value problem:

$$\begin{cases} -x''(t) = f(t, x(t), x'(t)), & t \in [0, \pi], \\ x(0) = x(\pi) = 0. \end{cases} \tag{E 5.4.1}$$

Put $X = C_0^1([0,\pi], R^n)$, $Y = L^1([0,\pi], R^n)$ and let $L : D(L) \subset X \to Y$ be defined by $Lx(\cdot) = x''(\cdot)$, where

$$D(L) = \{x \in X : x'(\cdot) \text{ is absolutely continuous in } [0,\pi],\ x''(\cdot) \in Y\}.$$

Then $Ker(L) = \{0\}$ and $Im(L) = Y$. Let a mapping $N : C_0^1([0,\pi], R^n) \to Y$ be defined by

$$(Nx)(t) = f(t, x(t), x'(t)) \quad \text{for all } t \in [0,\pi].$$

Then, by Lebesgue's Theorem, N is a continuous bounded mapping. Now, the problem (E 5.4.1) is equivalent to the following equation:

$$Lx = Nx, x \in D(L). \tag{E 5.4.2}$$

Since $L^{-1} : Y \to X_0$ is given by

$$(L^{-1}y)(t) = \frac{t}{\pi}\int_0^\pi \int_0^s y(l)dl - \int_0^t \int_0^s y(l)dl,$$

i.e.,

$$(L^{-1}y)(t) = \frac{1}{\pi}[\int_0^t s(\pi - t)y(s)ds + \int_x^\pi t(\pi - s)y(s)ds,$$

L^{-1} is continuous and compact, and consequently, it follows that N is L-compact on any bounded subset of X.

Theorem 5.4.1. Suppose that the following conditions hold:

(1) There exist $a, b > 0$ such that $a + b < 1$ and $g(\cdot) \in L^1([0,\pi], R_+)$ such that, for almost all $t \in [0,\pi]$,

$$(x, f(t,x,y)) \le a|x|^2 + b|x||y| + g(t)|x| \quad \text{for all } (x,y) \in R^n \times R^n;$$

(2) There exist $c \ge 0$ and $h(\cdot) \in L^1([0,\pi], R_+)$ such that, for all $x \in R^n$ with $|x| \le \pi(1 - a - b)^{-1}\|g\|_{L^1}$,

$$|f(t,x,y)| \le c|y|^2 + h(t) \quad \text{for almost all } t \in [0,\pi],\ y \in R^n.$$

Then the problem (E 5.4.1) has a solution.

Proof. Consider the family of problem:

$$\begin{cases} -x''(t) = \lambda f(t, x(t), x'(t)), & t \in [0,\pi], \\ x(0) = x(\pi) = 0. \end{cases} \tag{E 5.4.3}$$

Let x be a possible solution of (E 5.4.3) for some $\lambda \in (0,1)$. Then we have

$$\begin{aligned} -(x''(t), x(t)) &= \lambda(f(t, x(t), x'(t)), x'(t)) \\ &\le a|x(t)|^2 + b|x(t)||x'(t)| + g(t)|x(t)|. \end{aligned}$$

So, by integrating over $[0,\pi]$, it follows that

$$\|x'\|_{L^2}^2 \le a\|x\|_{L^2}^2 + b\|x\|_{L^2}\|x'\|_{L^2} + \|g\|_{L^1}\|x\|_0, \tag{5.4.1}$$

where $\|x\|_0 = max_{t\in[0,\pi]}\|x(t)\|$. Notice that $\|x\|_{L^2} \le \|x'\|_{L^2}$ and $\|x\|_0 \le \sqrt{\pi}\|x'\|_{L^2}$, so (5.4.1) implies that

$$\|x\|_0^2 \le \pi(1-a-b)^{-1}\|g\|_{L^1}\|x\|_0,$$

and thus we have

$$\|x\|_0 \le \pi(1-a-b)^{-1}\|g\|_{L^1} = r_1. \tag{5.4.2}$$

By (5.4.2) and (5.4.1), we get

$$\|x'\|_{L^2} \le \sqrt{\pi}(1-a-b)^{-1}\|g\|_{L^1} = r_2. \tag{5.4.3}$$

Now, from (E 5.4.3) and the assumption (2), we get

$$|x''(t)| \le c|x'(t)|^2 + h(t),$$

so $x \in D(L)$, and by (5.4.3) it follows that

$$\|x''\|_{L^1} \le c\pi(1-a-b)^{-2}\|g\|_{L^1}^2 + \|h\|_{L^1} = r_3. \tag{5.4.4}$$

On the other hand, by the boundary condition, there exists $s_i \in [0,\pi]$ such that $x_i'(s_i) = 0$ for $1 \le i \le n$ and thus

$$|x_i'(t)| \le |\int_{s_i}^t x_i''(s)ds| \le \|x''\|_{L^1},$$

which together with (5.4.4) imply that

$$\|x'\|_0 \le r_3. \tag{5.4.5}$$

Now, we take $r = 1 + r_1 + r_3$. Then $\|x\|_{C^1} \le r$ for all $\lambda \in (0,1)$ and every possible solution x of (E 5.4.3). Put $\Omega = B(0,r) \subset C_0^1([0,\pi], R^n)$. Then we have $Lx \ne tNx$ for all $(t,x) \in (0,1) \times \partial\Omega$. If $Lx = Nx$ for some $x \in \partial\Omega$, we know that the problem (E 5.4.1) has a solution. Otherwise, we have

$$Lx \ne tNx \quad \text{for all } (t,x) \in [0,1] \times \partial\Omega.$$

Thus $D_J(L-N,\Omega,0) = D_J(L,\Omega,0) \ne 0$, consequently, it follows that $Lx = Nx$ has a solution in $\overline{\Omega}$. Thus the problem (E 5.4.1) has a solution. This completes the proof.

In the following, let $X = C([0,1], R^n)$, $Y = L^1([0,1], R^n)$, $Q: Y \to Y$ be a mapping such that $Qy(\cdot) = \int_0^1 y(s)ds$ and $L: D(L) \subset C([0,1], R^n) \to Y$ be a mapping defined by

$$Lx(t) = x'(t) \quad \text{for all } x(\cdot) \in D(L),$$

where

$$D(L) = \{x(\cdot) \text{ is absolutely continuous on } [0,1] : x(0) = x(1)\}.$$

Then L is a Fredholm mapping of index zero.

Lemma 5.4.2. Let $r > 0$ and $v(\cdot) \in C^1(R^n, R)$ be such that $v'(x) \neq 0$ for $|x| = r$, where v' is the gradient of v, and let $V : X \to Z$ be defined by $V(x(\cdot))(t) = v'(x(t))$ for all $t \in [0,1]$. Then we have

$$D_J(L - V, B(0,r) \cap D(L), 0) = D_J(L - QV, B(0,r) \cap D(L), 0).$$

Proof. Consider the homotopy $T(\alpha, x)) = \alpha V(x(\cdot)) - (1-\alpha)QV(x(\cdot))$ for all $\alpha \in [0,1]$ and $x(\cdot) \in X$. We claim that $Lx(\cdot) \neq T(\alpha, x(\cdot))$ for all $\alpha \in [0,1]$ and $x(\cdot) \in X$. If not, there exist $\alpha \in [0,1]$ and $x(\cdot) \in X$ such that

$$Lx(t) \neq T(\alpha, x(t)) \quad \text{for all } t \in [0,1]. \tag{5.4.6}$$

Multiply both sides of (5.4.6) by $x'(t)$ and integrate over $[0,1]$, one gets

$$\int_0^1 |x'(s)|^2 ds = 0.$$

Thus $x(t)$ is a constant, $|x(t)| = r$ and $v(x(0)) = 0$, which is a contradiction. Thus it follows from Theorem 5.2.2 that

$$D_J(L - V, B(0,r) \cap D(L), 0) = D_J(L - QV, B(0,r) \cap D(L), 0).$$

This completes the proof.

Theorem 5.4.3. Suppose that the following conditions are satisfied:

(1) There exist $v(\cdot) \in C^1([0,1], R^n)$ such that $\lim_{|x|\to\infty} v(x) = +\infty$ and $\beta(\cdot) \in L^1([0,1], [0, +\infty))$ such that $(v'(x), f(t,x)) \leq \beta(t)$ for all $x \in R^n$ and almost all $t \in [0,1]$.

(2) There exist $r > 0$ and $w(\cdot) \in C^1(R^n, R)$ such that $(v'(x), w(x)) > 0$ for all x with $|x| \geq r$ and $\int_0^1 (w'(x(s)), f(s, x(s)) ds) \leq 0$ for all $x(\cdot) \in D(L)$ satisfying $\min_{t\in[0,1]} |x(t)| \geq r$.

Then the following equation:

$$\begin{cases} x'(t) = f(t, x(t)), & t \in [0,1], \\ x(0) = x(1), \end{cases} \tag{E 5.4.4}$$

has a solution.

Proof. First, we claim that there exists $r_0 > 0$ such that the solution $x_\lambda(\cdot)$ of the following equation:

$$\begin{cases} x'(t) = -(1-\lambda)v'(x(t)) + \lambda f(t, x(t)), & t \in [0,1], \\ x(0) = x(1), \end{cases} \qquad (E\ 5.4.5)$$

satisfies $\|x(\cdot)\|_0 \leq r_0$. If not, there exist a sequence $(\lambda_n) \subset [0,1]$ and a sequence $(x_{\lambda_n}(\cdot))$ of solutions such that $\|x_{\lambda_n}\|_0 \geq n$. By the assumption (1), we have

$$\frac{d}{dt}v(x_{\lambda_n}(t)) = -(1-\lambda_n)|v'(x_{\lambda_n}(t))|^2 + \lambda_n(v'(x_{\lambda_n}(t)), f(t, x_{\lambda_n}(t))) \leq \beta(t).$$

Extend $x_{\lambda_n}(\cdot)$ and $\beta(\cdot)$ to R with period 1 and then we have

$$v(x_{\lambda_n}(t)) \leq v'(x_{\lambda_n}(s)) + \int_s^t \beta(s)ds, s \in R \quad \text{for all } t \in [s, s+1].$$

Therefore, we have

$$\max_{t\in[0,1]} v(x_{\lambda_n}(t)) \leq \min_{t\in[0,1]} v(x_{\lambda_n}(t)) + \|\beta\|_1. \qquad (5.4.7)$$

Now, from $\|x_{\lambda_n}\|_0 \geq n$ and (5.4.7), we deduce $\min_{t\in[0,1]} v(x_{\lambda_n}(t)) \to \infty$, which implies that $\min_{t\in[0,1]} |x_{\lambda_n}(t)| \to \infty$. Thus there exists $N > 0$ such that, for $n \geq N$,

$$\min_{t\in[0,1]} |x_{\lambda_n}(t)| \geq r. \qquad (5.4.8)$$

Now, we have

$$\begin{aligned}\frac{d}{dt}w(x_{\lambda_n}(t)) &= -(1-\lambda_n)(v'(x_{\lambda_n}(t)), w'(x_{\lambda_n}(t))) \\ &\quad + \lambda_n(w'(x_{\lambda_n}(t)), f(t, x_{\lambda_n}(t))).\end{aligned}$$

From (5.4.9) and the assumption (2), one deduces that

$$0 = \int_0^1 \frac{d}{dt}w(x_{\lambda_n}(t))dt < 0,$$

which is a contradiction. Thus the claim is true. Choose $r_1 > \max\{r, r_0\}$. By Lemma 5.4.2, we have

$$|D_J(L - V, B(0,r) \cap D(L), 0)| = |D_J(L - QV, B(0,r) \cap D(L), 0)|.$$

By Proposition 5.2.4 and Theorem 1.2.15, we get

$$|D_J(L - V, B(0,r) \cap D(L), 0)| = 1.$$

Thus the problem (E 5.4.4) has a solution.

5.5 Exercises

1. Let X be a Banach space, $T : X \to X$ be a linear bounded Fredholm operator and $K : X \to X$ be a linear continuous compact mapping. Show that $T + K$ is a Fredholm mapping.

2. Assume that the conditions of Exercise 1 hold and $Ind(T) = 0$. Show that
$$Ind(T + K) = 0.$$

3. Let $L : D(L) \subset C([0,1], R^n) \to L^1([0,1]; R^n) \times R^n$ be defined by
$$Lx(t) = (x'(t), Mx(0) + Nx(1)) \quad \text{for all } x(\cdot) \in D(L),$$
where M, N are $n \times n$ real matrices. Show that L is a Fredholm mapping of index zero.

4. Let L be defined as in Exercise 3. Construct the projections P, Q such that $Im(P) = Ker(L)$ and $Im(L) = Ker(Q)$.

5. Let X, Y be real normed spaces, $L : D(L) \subset X \to Y$ be a Fredholm mapping of index zero and $A : X \to Y$ be a linear mapping such that A is L-compact on any bounded subset of A. Assume that $Ker(L-A) = \{0\}$. Show that
$$(L - A)(D(L)) = Y,$$
$$(L - A)^{-1} = (I - P - JQA - K_{PQ}A)^{-1}(JQA + K_{PQ}A).$$

6. Let $T : \overline{\Omega} \to Y$ be L-compact. Assume that $L - T$ is one to one on $D(L) \cap \overline{\Omega}$. Show that, for all $z \in (L - T)(D(L) \cap \Omega)$,
$$|D_J(L - T - z, D(L) \cap \Omega,)| = 1.$$

7. Let $T : \overline{\Omega} \to Y$ be L-compact. Suppose the following conditions hold:

   (1) $\|Lx - Tx\|^2 \geq \|Tx\|^2 - \|Lx\|^2$ for all $x \in (D(L) \setminus Ker(L)) \cap \partial\Omega$;

   (2) $Tx \notin Im(L) = 0$ for all $x \in Ker(L) \cap \partial\Omega$;

   (3) $deg(QT_{Ker(L)}, \Omega \cap Ker(L), 0) \neq 0$, where $Q : Y \to Y$ is the projection such that $Ker(Q) = Im(L)$.

   Show that $Lx = Tx$ has a solution in $D(L) \cap \overline{\Omega}$.

8. Let H be a Hilbert space, $T : \overline{\Omega} \to H$ L-compact. Suppose the following conditions hold:

   (1) $(Lx - Tx, Lx) \geq 0$ for all $x \in (D(L) \setminus Ker(L)) \cap \partial\Omega$;

(2) $Tx \notin Im(L) = 0$ for all $x \in Ker(L) \cap \partial\Omega$;

(3) $deg(QT_{Ker(L)}, \Omega \cap Ker(L), 0) \neq 0$, where $Q : H \to H$ is the projection such that $Ker(Q) = Im(L)$.

Show that $Lx = Tx$ has a solution in $D(L) \cap \overline{\Omega}$.

9. Let $f : [0, \pi] \times R^n \times R^n \to R^n$ be a function satisfying the Carathéodory condition. Assume that the following condtions hold:

(1) There exist $a, b \in R$ such that $a + b < 1$ and

$$(x - u, f(t, x, y) - f(t, u, v)) \leq a|x - u|^2 + b|x - u||y - u|$$

for all $x, y, u, v \in R^n$ and a.e. $t \in [0, \pi]$.

(2) There exist $c \geq 0$ and $h \in L^1([0, \pi], [0, +\infty)]$ such that, for all $x \in R^n$ with $|x| \leq \pi(1 - a - b)^{-1}\|f(t, 0, 0)\|_{L^1}$,

$$|f(t, x, y)| \leq c|y|^2 + h(t) \quad \text{for all } y \in R^n, \text{ and almost all } t \in [0, \pi].$$

Show that the following equation:

$$\begin{cases} -x''(t) = f(t, x(t), x'(t)), & t \in [0, \pi], \\ x(0) = x(\pi) = 0 \end{cases}$$

has a unique solution.

10. Let $f : [0, \pi] \times R^n \times R^n \to R^n$ be a function satisfying the Carathéodory condition. Assume that there exist $a, b \in R$ such that $a + b < 1$ and

$$|f(t, x, y) - f(t, u, v)| \leq a|x - u| + b|y - u|$$

for all $x, y, u, v \in R^n$ and almost all $t \in [0, \pi]$. Show that the following equation:

$$\begin{cases} -x''(t) = f(t, x(t), x'(t)), & t \in [0, \pi], \\ x(0) = x(\pi) = 0 \end{cases}$$

has a unique solution.

11. Let $f : [0, \pi] \times R^n \to R^n$ be a function satisfying the Carathéodory condition. Assume that there exist a number $a < 1$ and $g \in L^1([0, \pi], [0, +\infty)])$ such that

$$(x, f(t, x)) \leq a|x|^2 + g(t)|x| \quad \text{for all } x \in R^n, \text{ and alomst all} t \in [0, \pi].$$

Show that the following equation:

$$\begin{cases} -x''(t) = f(t, x(t)), & t \in [0, \pi], \\ x(0) = x(\pi) = 0 \end{cases}$$

has a solution.

12. Let $f : [0,1] \times R^n \times R^n \to R^n$ be a function satisfying the Carathéodory condition and $v \in C^1(R^n, R)$. Suppose that the following conditions are satisfied:

    (1) There exists $r > 0$ such that $(v'(x), f(t,x)) \leq 0$ for all $x \in R^n$ with $|x| \geq r$ and almost all $t \in [0,1]$;

    (2) $v'(x) \neq 0$ for all $x \in R^n$ with $|x| \geq r$ and $\lim_{|x|\to\infty} v(x) = +\infty$.

    Show that the following equation:

$$\begin{cases} x'(t) = f(t, x(t)), & t \in [0,1], \\ x(0) = x(1) \end{cases}$$

    has a solution.

13. Let $L : D(L) \subset X \to Y$ be a Fredholm mapping of index zero, $\Omega \subset X$ be an open bounded subset, $D(L) \cap \Omega \neq \emptyset$ and $T : \overline{\Omega} \to 2^Y$ be a mapping with closed convex values. Assume that QT and $K_{PQ}T$ are upper semicontinuous mapping such that $QT(\overline{\Omega})$ and $K_{PQ}T(\overline{\Omega})$ are relatively compact and $Lx \notin Tx$ for all $x \in \partial\Omega \cap D(L)$. Construct the coincidence degree for L and T on $\Omega \cap D(L)$.

14. Let $L : D(L) \subset X \to Y$ be a Fredholm mapping of index zero, $\Omega \subset X$ be an open bounded subset, $D(L) \cap \Omega \neq \emptyset$ and $T : \overline{\Omega} \to Y$ be a mapping such that QT and $K_{PQ}T$ are continuous countably condensing mapping and $Lx \neq Tx$ for all $x \in \partial\Omega \cap D(L)$. Construct the coincidence degree for L and T on $\Omega \cap D(L)$.

Chapter 6

DEGREE THEORY FOR MONOTONE-TYPE MAPS

Monotone-type mappings are a class of mappings without continuous and compact conditions. The concept of monotone mapping was introduced by Kachurovski, Vainberg, Zarantonello in 1960, and it plays a very important role in studying the weak solution of the partial differential equations in divergence form and variational inequality problems. Minty, Browder, Brézis, Rockafellar, Crandall, Gossez, etc., made significant contributions to monotone operator theory. It was shown by Skrypnik, Browder, Berkovitz, Mustonen, Kartsatos, and others that it is possible to construct the degree theory for monotone-type mappings.

The goal of this chapter is to introduce the degree theory for monotone-type mapping. Chapter 6 has seven sections.

In Section 6.1, we introduce some basic geometric properties of Banach spaces and various types of monotone and pseudomonotone maps and also (S_+), $(S_+)_{0,L}$ and L-(S_+)-mappings. Many examples and properties of these maps are presented in Section 6.1.

Section 6.2 presents the degree theory for monotone mappings of class (S_+).

In Section 6.3, using the results of Section 6.2, we present the degree theory for perturbations of maximal monotone mappings and various properties are also presented.

In Section 6.4, using the results of chapters 2, 3, we present the topological degree for multivalued mappings of class $(S_+)_{0,L}$. Some properties of this degree are presented in theorems 6.4.4, 6.4.5, and 6.4.6.

A degree for multivalued mappings of class L-(S_+) type is presented in Section 6.5 (here L is a Fredholm mapping of index zero type). Various properties are presented in Theorems 6.5.2, 6.5.3 and 6.5.5. The coincidence degree of L and a pseudomonotone mapping is also presented in this section.

Section 6.6 presents various results concerning the computation of the topological degree for a variety of mappings.

Section 6.7 gives various existence results for the partial differential equations and evolution equations.

6.1 Monotone Type-Mappings in Reflexive Banach Spaces

In this section, we introduce some monotone-type mappings and discuss their properties. We first recall some geometrical properties of Banach spaces.

Definition 6.1.1. Let X be a normed space, X^{**} be the second dual space of X and $K : X \to X^{**}$ be defined by $Kx(f) = f(x)$. If $KX = X^{**}$, then X is said to be reflexive.

Theorem 6.1.2. Let X be a normed space and X^* be the dual space of X. Then the following are equivalent:

(1) X is reflexive;

(2) X^* is reflexive;

(3) The closed unit ball of X is sequentially weak compact, i.e., each $(x_n)_{n=1}^{\infty}$ in the closed unit ball has a weakly convergent sequence;

(4) For all $f \in X^* \setminus \{0\}$, there exists $x \in X$ with $\|x\| = 1$ such that $f(x) = \|f\|$.

Definition 6.1.3. Let X be a Banach space. X is said to be strictly convex if, for any $x, y \in X$, $\|x\| = \|y\| = 1$ and $\|x + y\| = 2$ imply that $x = y$.

The following proposition follows directly from Definition 6.1.3.

Proposition 6.1.4. Let X be a Banach space. The following statements are equivalent:

(1) X is strictly convex;

(2) If, for any $x, y \in X$, $\|x\| = \|y\| = 1$ and $x \neq y$, then $\|x + y\| < 1$;

(3) Every point on the unit sphere is an extreme point;

(4) If $f \in X^*$ is nonzero and $\|x\| = \|y\| = 1$ such that $f(x) = f(y) = \|f\|$, then $x = y$.

Definition 6.1.5. A Banach space X is said to be locally uniform convex if, for any $x \in X$ with $\|x\| = 1$ and $\epsilon \in (0, 2]$, there exists $\delta(x) > 0$ such that, for any $y \in X$ with $\|y - x\| \geq \epsilon$, we have $\|x + y\| < 2 - \delta(x)$.

Definition 6.1.6. A Banach space X is said to be uniformly convex if, for any $\epsilon \in (0, 2]$, there exists $\delta > 0$ such that, for any $x, y \in X$ with $\|y - x\| \geq \epsilon$, we have $\|x + y\| < 2 - \delta$.

Some well-known uniformly convex spaces are Hilbert spaces, l^p, L^p, and the Sobolev space $W^{m,p}$, where $p > 1$ and $m > 0$ is an integer.

Proposition 6.1.7. A uniformly convex Banach space is reflexive.

In the sequel, let E be a real reflexive Banach space and E^* be the dual space of E, and $\rightharpoonup$ represents the weak convergence. By [287], we may assume that both E and E^* are locally uniformly convex. Let $J : E \to E^{**}$ be the duality mapping, i.e.,

$$\|Jx\| = \|x\|, \quad (Jx, x) = \|x\|^2.$$

Proposition 6.1.8. Let E be a real reflexive Banach space and $J : E \to E^{**}$ be the duality mapping. Then we have the following:

(1) If E^* is locally uniform convex, then J is single valued and continous;

(2) If E is locally uniform convex, $(x_n)_{n=1}^{\infty} \subset E$ is a sequence converging weakly to x_0 as $n \to \infty$, $\|x_n\| \to \|x_0\|$ and $Jx_n(x_0) \to \|x_0\|^2$, then $x_n \to x_0$.

Proof. (1) For all $x \neq 0$, if $f_1, f_2 \in Jx$, then we have $(f_1 + f_2)(x) = 2\|x\|^2$ and thus $\|f_1 + f_2\| = 2\|x\|$. In addition, $\|f_1\| = \|f_2\| = \|x\|$, and so it follows from the locally uniform convexity of E^* that $f_1 = f_2$.

Next, assume that $x_n \to x_0$ in E and we may assume also that $(Jx_n))$ has a weakly convergent sequence (Jx_{n_k}) with $Jx_{n_k} \rightharpoonup f_0$ by reflexivity of E^*. Then we have

$$Jx_{n_k}(x_{n_k}) \to \|x_0\|^2, \quad Jx_{n_k}(x_0) \to f_0(x_0).$$

In addition, we have

$$\|J(x_{n_k})(x_{n_k}) - J(x_{n_k})(x_0)\| = 0.$$

Therefore, we get $f_0(x_0) = \|x_0\|^2$ and, from this, we deduce that

$$\|f_0\| = \|x_0\|, \quad \|J(x_{n_k})\| \to \|x_0\|, \quad \|J(x_{n_k}) + f_0\| \to \|x_0\| = \|f_0\|.$$

From the locally uniform convexity of E^*, we deduce $J(x_{n_k}) \to Jx_0$.

(2) For simplicity, we may assume that $Jx_n \rightharpoonup f_0$ by taking a subsequence since E^* is reflexive. By assumption, we have

$$Jx_n(x_n + x_0) = \|x_n\|^2 + Jx_n(x_0) \to 2\|x_0\|^2$$

and so $\|x_n + x_0\| \to \|x_0\|$. From the locally uniformly convexity of E, we deduce $x_n \to x_0$. This completes the proof.

The following is the well-known Mazur's separation theorem for convex sets:

Theorem 6.1.9. Let X be a Banach space, C_1 be an compact convex set and C_2 be a closed convex set such that $C_1 \cap C_2 = \emptyset$. Then there exists a $f \in X^* \setminus \{0\}$ such that

$$\sup_{x \in C_1} f(x) < \inf_{x \in C_2} f(x).$$

Next, we recall some terminology as follows:

Definition 6.1.10. Let E be a real reflexive Banach space and E^* be the dual space of E.

(1) An operator $T : D(T) \subseteq E \to 2^{E^*}$ is said to be monotone if

$$(f - g, x - y) \geq 0$$

for all $x, y \in D(T)$, $f \in Tx$ and $g \in Ty$;

(2) T is said to be maximal monotone if T is monotone and does not have a proper monotone extension.

Note that when E is a Hilbert space the mapping T is said to be pseudo-contractive if $I - T$ is maximal monotone (see [51]).

As a consequence of Zorn's Lemma, every monotone mapping has a maximal monotone extension. We leave the details to the reader.

Definition 6.1.11. Let E be a reflexive Banach space. A multi-valued operator $T : D(T) \subseteq E \to 2^{E^*}$ is said to be a mapping of class (S_+) if it satisfies the following conditions:

(1) Tx is bounded closed and convex for each $x \in D(T)$;

(2) T is weakly upper semicontinuous in each finite dimensional space, i.e., for each finite dimensional space F with $F \cap D(T) \neq \emptyset$, $T : F \cap D(T) \to 2^{E^*}$ is upper semicontinuous in the weak topology;

(3) if $\{x_n\} \subset D(T)$ and $\{x_n\}$ converges weakly to x_0 in E such that

$$\limsup_{n\to\infty} (f_n, x_n - x_0) \leq 0 \quad \text{for some } f_n \in Tx_n,$$

then $x_n \to x_0 \in D(T)$ and $\{f_n\}$ has a subsequence which converges weakly to $f_0 \in Tx_0$ in E^*.

When E is a Hilbert space, we say that T is pseudocompact if $I - T$ is a mapping of class (S_+) (see [51]). One may easily see that a compact mapping is pseudocompact.

Definition 6.1.12. Let E be a reflexive Banach space. A family of operators $\{T_t : D(T_t) \subseteq E \to 2^{E^*}\}_{t\in[0,1]}$ is said to be a homotopy of mappings of class (S_+) if T_t satisfies the conditions (1), (2) in Definition 6.1.11 for each $t \in [0, T]$ and the following condition:

(3)' If $t_n \to t_0$ and $x_n \in D(T_{t_n})$, $\{x_n\}$ converges weakly to x_0 in E such that

$$\limsup_{n\to\infty} (f_n, x_n - x_0) \leq 0 \quad \text{for some } f_n \in T_{t_n} x_n,$$

then $x_n \to x_0 \in D(T_{t_0})$ and $\{f_n\}$ has a subsequence converging weakly to some $f_0 \in T_{t_0}x_0$ in E^*.

Definition 6.1.13. Let $T : D(T) \subseteq E \to 2^{E^*}$ be a mapping satisfying the conditions (1), (2) in Definition 6.1.11. Let $(x_j) \subset D(T)$, $x_j \rightharpoonup x_0 \in D(T)$ and $f_j \in Tx_j$.

(1) If $\limsup_{j\to\infty}(f_j, x_j - x_0) \leq 0$ implies that

$$(f_0, x_0 - v) \leq \liminf_{j\to\infty}(f_j, x_j - v) \quad \text{for all } v \in D(T),\ f_0 \in Tx_0,$$

then T is said to be a pseudomonotone mapping;

Let $(x_j) \subset D(T)$, $x_j \rightharpoonup x_0 \in D(T)$, $f_j \in Tx_j$ and $f_j \rightharpoonup f_0$.

(2) If $\limsup_{j\to\infty}(f_j, x_j - x_0) \leq 0$ implies that

$$f_0 \in Tx_0, \quad (f_0, x_0) = \lim_{j\to\infty}(f_j, x_j),$$

then T is said to be a generalized pseudomonotone mapping.

Lemma 6.1.14. If $J : E \to E^*$ is the duality mapping, then J is a continuous mapping of class (S_+) and J is also monotone.

Proof. Let $x_n \to x_0$. We may assume that $Jx_n \rightharpoonup f_0 \in E^*$. Since $\lim_{n\to\infty}(Jx_n, x_n - x_0) = 0$, we have $(f_0, x_0) = \|x_0\|^2$. Therefore, we get $\|Jx_n\| \to \|x_0\| = \|f_0\|$. The local uniform convexity of E^* implies that $Jx_n \to Jx_0$, and so J is continuous and, furthermore, J monotone is obvious.

Next, we prove that J is a mapping of class (S_+). Let

$$x_n \rightharpoonup x_0, \quad \limsup_{n\to\infty}(Jx_n, x_n - x_0) \leq 0.$$

We may assume that $Jx_n \rightharpoonup f_0 \in E^*$. Since

$$\|x_0\|^2 \leq \liminf_{n\to\infty}(Jx_n, x_n) \leq (f_0, x_0),$$

we have $\|x_n\| \to \|x_0\|$, $\|f_0\| = \|x_0\|$. Thus by Proposition 6.1.8, $x_n \to x_0$ and $Jx_n \to Jx_0$. This completes the proof.

Definition 6.1.15 [163] Let E be a separable reflexive Banach space and L be a dense subspace of E. A mapping $T : D(T) \subset E \to E^*$ is said to be a mapping of class $(S_+)_L$ if, for any sequence of finite dimensional subspaces F_j of L with $\overline{\cup_{j=1}^{\infty} F_j} = E$, $h \in E^*$, $\{u_j\}_{j=1}^{\infty} \subset D(T)$ with $u_j \rightharpoonup u_0$ and

$$\limsup_{j\to\infty}(Tu_j - h, u_j) \leq 0, \quad \lim_{j\to\infty}(Tu_j - h, v) = 0$$

for all $v \in \cup_{j=1}^{\infty} F_j$, we have $u_j \to u_0$, and $u_0 \in D(T)$, $Tu_0 = h$. If $h = 0$, then we call T a mapping of class $(S_+)_{0,L}$.

Definition 6.1.16. Let E be a reflexive Banach space and L be a subspace of E. A multi-valued mapping $T : D(T) \subseteq E \to 2^{E^*}$ is said to be a mapping of class $(S_+)_L$ if it satisfies the following conditions:

(i) Tx is bounded closed and convex for each $x \in D(T)$;

(ii) T is weakly upper semicontinuous in each finite dimensional space, i.e., for each finite dimensional space F of L with $F \cap D(T) \neq \emptyset$, $T : F \cap D(T) \to 2^{E^*}$ is upper semicontinuous in the weak topology;

(iii) if, for any sequence of finite dimensional subspaces F_j of L with $L \subseteq \overline{\cup_{j=1}^{\infty} F_j}$, $h \in E^*$, $\{x_j\}_{j=1}^{\infty} \subset D(T) \cap L$ and $x_j \rightharpoonup x_0$ such that

$$\limsup_{j\to\infty}(f_j - h, x_j) \le 0, \quad \lim_{j\to\infty}(f_j - h, v) = 0$$

for all $v \in \cup_{j=1}^{\infty} F_j$ and some $f_j \in Tx_j$, then $x_j \to x_0 \in D(T)$ and $h \in Tx_0$. If $h = 0$, then we call T a mapping of class $(S_+)_{0,L}$.

Remark. If T is a mapping of class $(S_+)_L$, then, for any $p \in E^*$, $T - p$ is a mapping of class $(S_+)_{0,L}$.

When E is a Hilbert space, we say that T is L-pseudocompact if $I - T$ is a mapping of class $(S_+)_L$, and T is L_0-pseudocompact if $I - T$ is a mapping of class $(S_+)_{0,L}$.

Definition 6.1.17. Let E be a reflexive Banach space and let L be a subspace of E. A family of mappings $\{T_t : D(T_t) \subseteq E \to 2^{E^*}\}_{t\in[0,1]}$ is called a homotopy of mappings of class $(S_+)_L$ if the conditions (i) and (ii) in Definition 6.1.16 hold for each $t \in [0, T]$ and the following condition holds:

(iii) If, for any sequence of finite dimensional subspaces F_j of L with $L \subseteq \overline{\cup_{j=1}^{\infty} F_j}$, $h \in E^*$, $t_j \to t_0$, $x_j \in D(T_{t_j}) \cap L$ and $x_j \rightharpoonup x_0$ such that

$$\limsup_{j\to\infty}(f_j - h, x_j) \le 0, \quad \lim_{j\to\infty}(f_j - h, v) = 0,$$

for all $v \in \cup_{j=1}^{\infty} F_j$ and some $f_j \in T_{t_j}x_j$, then $x_j \to x_0 \in D(T_{t_0})$, $h \in T_{t_0}x_0$. If $h = 0$, then we call $\{T_t : D(T_t)\}_{t\in[0,1]}$ a homotopy of mappings of class $(S_+)_{0,L}$.

Proposition 6.1.18. If L is dense in E, then the duality mapping J is a mapping of class $(S_+)_L$.

Proof. Let $(F_j)_{j=1}^{\infty}$ be a sequence of finite dimensional subspaces of L with $\overline{\cup_{j=1}^{\infty} F_j} = E$, $h \in E^*$, $x_j \in L$ and $x_j \rightharpoonup x_0$ such that

$$\limsup_{j\to\infty}\langle Jx_j - h, x_j\rangle \le 0, \quad \lim_{j\to\infty}(Jx_j - h, v) = 0$$

for all $v \in \cup_{j=1}^{\infty} F_j$. Without loss of generality, we may assume that $Jx_j \rightharpoonup f_0$ in E^*. Then we have

$$\|x_0\|^2 \leq \limsup_{j\to\infty}(Jx_j, x_j) \leq (h, x_0), \quad (f_0 - h, v) = 0$$

for all $v \in \cup_{j=1}^{\infty} F_j$. But $\cup_{j=1}^{\infty} F_j$ is dense in E, so we have $f_0 = h$. Therefore it follows that

$$\|x_0\|^2 \leq \limsup_{j\to\infty}(Jx_j, x_j) \leq (h, x_0) \leq \|x_0\|^2.$$

Thus $\|x_j\| \to \|x_0\|$ as $j \to \infty$. This and the local uniform convexity of E together with $x_j \rightharpoonup x_0$ imply that $x_j \to x_0$ and $Jx_0 = h$. Therefore, J is a mapping of class $(S_+)_L$.

Now, assume that L is a Fredholm mapping of index zero type. Then there exist two linear continuous projections $P : H \to H$ and $Q : H \to H$ such that

$$Im(P) = Ker(L), \quad Ker(Q) = Im(L).$$

Also, we have

$$H = Ker(L) \oplus Ker(P), \quad H = Im(L) \oplus Im(Q)$$

as the topological direct sums.

Obviously, the restriction of L_P of L to $D(L) \cap Ker(P)$ is one to one and onto $Im(L)$, so its inverse $K_P : Im(L) \to D(L) \cap Ker(P)$ is defined. We denote by $K_{PQ} : H \to D(L) \cap Ker(P)$ the generalized inverse of L defined by $K_{PQ} = K_P(I - Q)$. Let $J : Im(Q) \to Ker(L)$ be a linear homeomorphism.

Definition 6.1.19. Let H be real Hilbert space, $L : D(L) \subseteq H \to H$ be a Fredholm mapping of index zero type and $T : D(T) \subseteq H \to 2^H$ be a set-valued mapping.

(1) If $I - P - (JQ + K_{PQ})T$ is maximal monotone mapping, then T is said to be L-maximal monotone;

(2) if $I - P - (JQ + K_{PQ})T$ is pseudomonotone (respectively, generalized pseudo monotone), then T is L-pseudomonotone (respectively, generalized L-pseudomonotone). If $I - P - (JQ + K_{PQ})T$ is also bounded, then T is said to be bounded L-pseudomonotone;

(3) If $I - P - (JQ + K_{PQ})T$ is a mapping of class (S_+), then T is called a mapping of class L-(S_+).

Remark. (1) If T is a mapping of class (S_+), $L = 0$, $P = I$, $Q = I$, and $J = -I$, then T is a mapping of class 0-(S_+).

(2) If $I-T$ is a mapping of class (S_+), in this case, T is called pseudocompact in [51], and $L = I$, $P = 0$, $Q = 0$, $J = I$, then T is a mapping of class I-(S_+). As a special consequence, if T is upper semicontinuous compact mapping with bounded closed convex values, then $I - T$ is a mapping of class (S_+), so T is a mapping of class I-(S_+).

(3) One may also easily see that, if T is L-compact, then T is a mapping of class L-(S_+).

Proposition 6.1.20. Let H be real Hilbert space, $L : D(L) \subset H \to H$ be a generalized Fredholm mapping of index zero type and $T_i : D(T_i) \subseteq H \to 2^H$ be a mapping of class L-(S_+) for $i = 1, 2$. Then $tT_1 + (1-t)T_2$ is a mapping of class L-(S_+) on $D(T_1) \cap D(T_2)$ for all $t \in [0, T]$.

Now, we give some examples of monotone-type mappings.

Example 6.1.21. Let $\Omega \subset R^n$ be an open bounded subset with smooth boundary. Assume that a_i, $b_i : R \to [0, +\infty)$ are continuous functions for $i = 1, 2, \cdots, n$. Suppose the following conditions are satisfied:

(1) $c_1 \leq a_i(x) \leq c_2$ for all $(t, x) \in R^2$, where $c_1, c_2 > 0$ are constants;

(2) $\Sigma_i[b_i(x_i) - b_i(y_i)](x_i - y_i) \geq 0$, where $x = (x_i), y = (y_i) \in R^N$;

(3) $|b_i(x)| \leq \beta|x| + \gamma$ for all $x \in R$ and $i = 1, 2, \cdots, n$;

(4) $\Sigma_i b_i(x_i)x_i \geq \alpha|x|^2 - c_0$ for all $x = (x_i) \in R^n$.

Let $A : L^2(\Omega) \times H_0^1(\Omega) \to H^*$ be defined as follows:

$$(A(u, v), w) = \int_\Omega [\Sigma_{i=1}^n a_i(u) b_i(D_i v) D_i w] dx$$

for all $u, w \in H_0^1(\Omega)$. Then we have the following:

(a) for each $v \in L^2(\Omega)$,

$$(A(v, u), w) \leq c_2 \beta \sqrt{\int_\Omega (\Sigma_{i=1}^n |D_i u|^2) dx} \sqrt{\int_\Omega (\Sigma_{i=1}^n |D_i w|^2) dx}$$

for all $u, w \in H_0^1(\Omega)$,

(b) $(A(v, u_1) - A(v, u_2), u_1 - u_2) \geq 0$ for $t \in R$, $u_1, u_2, \in H_0^1(\Omega)$ and $v \in L^2(\Omega)$ and $A(v, \cdot)$ is continuous and monotone for all $v \in R \times L^2(\Omega)$ and so it is also pseudomonotone.

A special case of Example 6.1.21 is that, if $a_i(x) = 1$ and $b_i(x) = x$ for all $x \in R$, then we get the Laplace operator

$$(-\Delta u, v) = \int_\Omega (\nabla u, \nabla v) dx \quad \text{for all } u, v \in H_0^1(\Omega).$$

Example 6.1.22. Let Ω be as in Example 6.1.21 and let $f(x,y) : R^2 \to R$ be a continuous function satisfying

$$|f(x,y)| \leq M|y| + g(x) \quad \text{for all } (x,y) \in R^2,$$

where $M > 0$ is a constant and $g(\cdot) \in L^2(\Omega)$. Then the following mapping A defined by

$$(Au, v) = \int_\Omega [(\nabla u, \nabla v) + f(x,u)v]dx \quad \text{for all } u, v \in H_0^1(\Omega)$$

is a mapping of class (S_+) and it is also pseudomonotone.

Example 6.1.23. Let Ω be as in Example 6.1.17, $p > 1$, and $a_\alpha : \Omega \times R^n \to R$ be a Caratheodory function, i.e., $a_\alpha(x, \xi)$ is measurable in x and continuous in ξ, where $\alpha = (\alpha_1, \alpha_2, \cdots, \alpha_N)$ is a multi-index. Assume that

$$|a_\alpha(x, \xi)| \leq C|\xi|^{p-1} + g(x)$$

for almost everywhere $x \in \Omega$, where $\xi = (\xi_\alpha : |\alpha| \leq m\}$, $C > 0$ is a constant, $g(\cdot) \in L^q$ and $p^{-1} + q^{-1} = 1$.

We consider the following partial differential equation:

$$\begin{cases} \Sigma_{|\alpha| \leq m}(-1)^{|\alpha|} D^\alpha a_\alpha(x, u, Du, \cdots, D^m u) + f(x,u) = 0, & x \in \Omega, \\ u(x) = 0, & x \in \partial\Omega, \end{cases}$$

and we define a mapping $A : H_0^m(\Omega) \to (H_0^m(\Omega)^*$ by

$$(Au, v) = \int_\Omega [\Sigma_{|\alpha| \leq m} a_\alpha(x, u, Du, \cdots, D^m u) D^\alpha v + f(x,u)v]dx$$

for all $u, v \in H_0^m(\Omega)$. By imposing suitable conditions on a_α, the above mapping A will be maximal monotone, pseudomonotone or of class (S_+).

Example 6.1.24. Consider the following wave equation with discontinuity:

$$\begin{cases} u_{tt}(t,x) - u_{xx}(t,x) + g(u(t,x)) + h(u(t,x)) = f(t,x), \\ t \in (0, 2\pi), x \in (0, \pi), \\ u(t,0) = u(t,\pi) = 0, \quad t \in (0, 2\pi), \\ u(0,x) = u(2\pi, x), \qquad x \in (0, \pi), \end{cases}$$

where $g : R \to R$ is a nondecreasing function with

$$|g(u)| \leq \alpha|u| + \beta \quad \text{for all } u \in R,$$

$h : R \to R$ is a continuous function satisfying

$$|h(u)| \leq \delta|u| + \gamma$$

and $f(\cdot) \in L^2((0, 2\pi) \times (0, \pi))$, where $\alpha > 0, \beta > 0, \delta > 0, \gamma > 0$ are constants.

Set

$$g_-(u) = \liminf_{s \to u} g(s), \quad g_+(u) = \limsup_{s \to u} g(s),$$

$$Gu = [g_-(u), g_+(u)].$$

Then $G : R \to 2^R$ is a maximal monotone mapping. Let a mapping $N : L^2((0, 2\pi) \times (0, \pi)) \to L^2((0, 2\pi) \times (0, \pi))$ be defined by

$$N(u(t, x)) = \{v(t, x) \in L^2((0, 2\pi) \times (0, \pi)) : v(t, x) \in Gu(t, x)\}$$

for all $u(t, x) \in L^2((0, 2\pi) \times (0, \pi))$.

From our assumption on g, we know that $D(N) = L^2((0, 2\pi) \times (0, \pi))$ and N is also maximal monotone, so N is upper semicontinuous from the strong topology of $L^2((0, 2\pi) \times (0, \pi))$ to the weak topology of $L^2((0, 2\pi) \times (0, \pi))$.

Let $L : D(L) \subset L^2((0, 2\pi) \times (0, \pi)) \to L^2((0, 2\pi) \times (0, \pi))$ be the wave operator $Lu = u_{tt} - u_{xx}$. Then it is well known that L is self-adjoint, densely defined and closed and $Ker(L)$ is infinite dimensional with $Ker(L)^{\perp} = Im(L)$. Thus L is a Fredholm mapping of zero index type and also the right inverse of L denoted by $L^{-1} : Im(L) \to Im(L)$ is compact.

Let $P : L^2((0, 2\pi) \times (0, \pi) \to Ker(L)$ be a projection. We assume that $\liminf_{i \to \infty}(P(f_j + h(u_j)), u_j - u_0) \geq 0$ for any $u_j \rightharpoonup u_0$ in $L^2((0, 2\pi) \times (0, \pi))$ and $f_j \in Nu_j$ (for example, take $g(x) = x$ for $x < 0$, $g(x) = x + 1$ for $x \geq 0$ and $h(x) = xsinx$ for $x \in R$). Then $N(u) + h(u)$ is L-pseudomonotone. In fact, if $u_j \rightharpoonup u_0$ in $L^2((0, 2\pi) \times (0, \pi))$, $f_j \in Nu_j$ with $f_j \rightharpoonup f_0$, $h(u_j) \rightharpoonup h_0$ and

$$\limsup_{j \to \infty}((I - P)u_j + P(f_j + h(u_j)) + L^{-1}(I - P)(f_j + h(u_j)), u_j - u_0) \leq 0,$$

then, since

$$\limsup_{j \to \infty}(P(f_j + h(u_j)), u_j - u_0) \geq 0$$

and L^{-1} is compact, we have

$$\limsup_{j \to \infty}(L^{-1}(I - P)(f_j + h(u_j)), u_j - u_0) = 0$$

and so $\limsup_{j \to \infty}((I - P)u_j, u_j - u_0) \leq 0$. Thus $(I - P)u_j \to (I - P)u_0$. Consequently, we get

$$\begin{aligned} &\lim_{j \to \infty}((I - P)u_j + P(f_j + h(u_j)) + L^{-1}(I - P)(f_j + h(u_j)), u_j) \\ &= ((I - P)u_0 + P(f_0 + h_0) + L^{-1}(I - P)(f_0 + h_0), u_0). \end{aligned}$$

Thus $N + h$ is L-pseudomonotone.

Lemma 6.1.25. Let X be a real Banach space and $T : D(T) \subset E \to 2^{E^*}$ be a monotone mapping. Then T is locally bounded on the interior of $D(T)$.

Proof. Let $x_0 \in int(D(T))$. Without loss of generality, we assume that $x_0 = 0$. There exists $r > 0$ such that $\overline{B_r(0)} \subset D(T)$. By the monotonicity of T, we have

$$(f - f_0, x - z) \geq 0$$

where $f_0 \in Tz$ is given and $x \in D(T), f \in Tx$. For each $n \geq 1$, let

$$M_n = \{z \in \overline{B_r(0)} : (f, x - z) \geq -n,\ x \in \overline{B_r(0)},\ f \in Tx\}.$$

Then we have

$$\overline{B_r(0)} = \cup_{n=1}^{\infty} M_n.$$

By the Baire's category theorem, there exists n_0 such that M_{n_0} has nonempty interior and so there exist $z_0 \in \overline{B_r(0)}$, $r_0 > 0$ such that $B_{r_0}(z_0) \subset M_{n_0}$. Since $-z_0 \in \overline{B_r(0)}$, there exists $m > 0$ such that

$$(f, x + z_0) \geq m \quad \text{for all } x \in \overline{B_r(0)},\ f \in Tx.$$

Therefore, we have

$$(f, 2x - y) \geq -(n_0 + m_0) \quad \text{for all } y \in \overline{B_{r_0}(0)}.$$

Now, for all $x \in \overline{B_{\frac{r}{4}}(0)}$, $f \in Tx$ and $z \in \overline{B_{\frac{r_0}{2}}(0)}$, we have

$$(f, z) = (f, 2x - (2x - z)) \geq -(n_0 + m_0)$$

and so

$$\|f\| \leq \frac{2}{r}(n_0 + m_0).$$

This completes the proof.

Proposition 6.1.26. Let $P : E \to 2^{E^*}$ be a bounded pseudomonotone mapping and $\{x_n\} \subset E$ be such that $x_n \rightharpoonup x_0$. If $f_n \in Px_n$ such that $f_n \rightharpoonup f_0$ and $\limsup_{n\to\infty}(f_n, x_n - x_0) \leq 0$, then $f_0 \in Px_0$ and $(f_n, x_n) \to (f_0, x_0)$.

Proof. Since $\limsup_{n\to\infty}(f_n, x_n - x_0) \leq 0$, we have

$$(f, x_0 - v) \leq \liminf_{n\to\infty}(f_n, x_n - v) \quad \text{for all } v \in E,\ f \in Px_0. \tag{6.1.1}$$

Putting $v = x_0$ in (6.1.1), it follows that

$$\liminf_{n\to\infty}(f_n, x_n - x_0) \geq 0$$

and so we have $(f_n, x_n) \to (f_0, x_0)$. Now, (6.1.1) becomes

$$(f, x_0 - v) \leq (f_0, x_0 - v) \quad \text{for all } v \in E,\ f \in Px_0. \tag{6.1.2}$$

But since Px_0 is bounded, closed and convex, by Mazur's separation theorem of convex subsets, we get $f_0 \in Px_0$. This completes the proof.

Lemma 6.1.27. Let E be a real reflexive Banach space, $T : D(T) \subset E \to 2^{E^*}$ be a maximal monotone mapping and let $P : E \to E^*$ be a bounded, coercive and demicontinuous pseudomonotone mapping. Then there exists $x_0 \in C$ such that

$$(f + Px_0, x - x_0) \geq 0 \quad \text{for all } x \in D(T),\ f \in Tx.$$

Proof. We first prove that, for any finite dimensional subspace F of E such that $F \cap D(T) \neq \emptyset$, there exists $x_F \in F$ such that

$$(f + Px_F, x - x_F) \geq 0 \text{ for all } x \in F \cap D(T),\ f \in Tx. \tag{6.1.3}$$

Since P is coercive, we notice that if the above conclusion is true, x_F must be in a bounded ball $B(0, R)$. Therefore, we may first prove that the above conclusion is true in the case that $D(T)$ is bounded. For the unbounded case, we find $x_n \in D(T)$ such that

$$(f + Px_n, x - x_n) \geq 0$$

for all $x \in F \cap D(T) \cap B(0, n)$ and $f \in Tx$. But since $\{x_n\}$ is bounded, we may assume that $x_n \to x_0$ as $n \to \infty$ and then we can use the demicontinuity of P to conclude that x_0 is the desired point.

In the following, we assume that $D(T)$ is bounded and suppose that the conclusion is not true. Then, for any $x \in F \cap D(T)$, there exist $z \in F \cap D(T)$ and $f_z \in Tz$ such that

$$(f_z + Px, z - x) < 0. \tag{6.1.4}$$

Now, if we take a compact set C such that $F \cap D(T) \subset C$, then $C = \cup_{z \in D(T)} U_z$, where $U_z = \{x \in F : (f_z + Px, z - x) < 0 \text{ for some } f_z \in Tz\}$. Notice that each U_z is open and so there exist finite many $z_i \in F \cap D(T)$ for $i = 1, 2, \cdots, N$ such that

$$C = \cup_{i=1}^{N} U_{z_i}.$$

Let $\{\alpha_1, \alpha_2, \cdots, \alpha_N\}$ be a partition of unity subordinated to the covering $\{U_{z_i}\}$. Then we define a mapping $K : C \to C$ by

$$Kx = \Sigma_{i=1}^{N} \alpha_i(x) z_i \quad \text{for all } x \in C.$$

Then K has a fixed point $x_0 \in C$, i.e., $x_0 = \Sigma_{i=1}^{N} \alpha_i(x_0) z_i$.

We may assume that $\alpha_i(x_0) > 0$ for $i = 1, 2, \cdots, N$. Otherwise, we exclude it. Now, we have

$$\begin{aligned} 0 &= (Kx_0 - x_0, \Sigma_{i=1}^{N} \alpha_i(x) f_{z_i} + Px_0) \\ &= \Sigma_{i,j=1}^{n} \alpha_i(x_0) \alpha_j(x_0) (f_{z_j} + Px_0, z_j - x_0) \\ &< 0, \end{aligned}$$

which is a contradiction. Thus (6.1.3) is true.

Finally, for each finite dimensional subspace F of E, we put

$$W_F = \{(x_F, Gx_F) : (x_F, Gx_F) \text{ satisfies } (6.1.3)\}.$$

Obviously, we have

$$\cap_{i=1}^n W_{F_i} \neq \emptyset.$$

If we denote by $\overline{W_F}^*$ the weak closure of W_F in E, then we have

$$\cap_{F \subset E, dim F < \infty} \overline{W_F}^* \neq \emptyset.$$

Take $(x_0, g_0) \in \cap_{F \subset E, dim F < \infty} \overline{W_F}^*$ and, for any $x \in D(T) \cap C$, a finite dimensional subspace F of E such that x, $x_0 \in F$. Then there exists $\{(x_j, Px_j)\} \subset W_F$ such that

$$x_j \rightharpoonup x_0, \quad Px_j \rightharpoonup g_0 \quad \text{as } j \to \infty.$$

Therefore, we have

$$(f + Px_j, x - x_j) \geq 0 \quad \text{for all } x \in F \cap D(T),\ f \in Tx.$$

Since P is pseudomonotone and C is closed and convex, it follows from Proposition 6.1.21 that

$$\limsup_{j \to \infty} (Px_j, x_j) \geq (g_0, x_0)$$

and so we have

$$(f + g_0, x - x_0) \geq 0 \quad \text{for all } x \in F \cap D(T),\ f \in Tx.$$

But since F is arbitrary, $(f + g_0, x - x_0) \geq 0$ for all $x \in D(T)$ and $f \in Tx$. Hence $x_0 \in D(T)$, which implies that

$$\limsup_{j \to \infty} (Px_j, x_j - x_0) \leq 0.$$

Thus we have $(Px_j, x_j) \to (Px_0, x_0)$ and so

$$(f + Px_0, x - x_0) \geq 0 \quad \text{for all } x \in D(T),\ f \in Tx.$$

This completes the proof.

Let $T : D(T) \subseteq E \to 2^{E^*}$ be a maximal monotone operator. Let $T_\lambda = (\lambda J^{-1} + T^{-1})^{-1}$ denote the Yosida approximation and $R_\lambda = I - \lambda J^{-1} T_\lambda$ the resolvent with respect to T_λ, respectively.

As a direct consequence of Lemma 6.1.27, we get the following:

Proposition 6.1.28. If $T : D(T) \to 2^{E^*}$ is a monotone mapping, then T is maximal monotone if and only if $T + \epsilon J$ is surjective for any $\epsilon > 0$.

Proof. We assume that both the spaces E and E^* are locally uniform convex. If T is maximal monotone, then, for any $p^* \in E^*$, $J - p^*$ is a continuous bounded coercive monotone mapping and, thus, pseudomonotone. By Lemma 6.1.27, there exists $x_0 \in E$ such that

$$(f + \epsilon Jx_0 - p^*, x - x_0) \geq 0 \quad \text{for all } x \in D(T),\ f \in Tx.$$

So $x_0 \in Tx_0$ and $-\epsilon Jx_0 + p^* \in Tx_0$, i.e., $p^* \in Tx_0 + \epsilon Jx_0$.

Conversely, suppose that $(f - g, x - x_0) \geq 0$ for all $x \in D(T)$ and $f \in Tx$. Then there exist $y_0 \in D(T)$ and $f_0 \in Ty_0$ such that $f_0 - Jy_0 = Jx_0 + g$. Therefore, it follows that

$$(Jx_0 - Jy_0, y_0 - x_0) \geq 0.$$

So we must have $x_0 = y_0$. This completes the proof.

Lemma 6.1.29. Let $T : D(T) \subseteq E \to E^*$ be a maximal monotone mapping. If $x \in D(T)$, then

$$\lim_{\lambda \to 0+} R_\lambda x = x, \qquad \lim_{\lambda \to 0+} T_\lambda x = f,$$

where $f \in Tx$ and $\|f\| = \min\{\|g\|, g \in Tx\}$.

Proof. By the monotonicity of T, it follows that

$$\|R_\lambda x\|^2 \leq -\lambda(R_\lambda x - y, g) - (x - y, T_\lambda x) \tag{6.1.5}$$

for all $x, y \in D(T)$ and $g \in Ty$. Put $y = x$ in (6.1.5) and then we have

$$\|T_\lambda x\| \leq \min\{\|g\| : g \in Tx\} = \|f\|. \tag{6.1.6}$$

By letting $\lambda \to 0^+$, it immediately yields that $R_\lambda x \to x$. Without loss of generality, we may assume that $T_\lambda x \rightharpoonup f_0$ as $\lambda \to 0^+$. Then the maximal monotonicity of T yields that $f_0 \in Tx$, and so from (6.1.6) and the locally uniform convexity of E^* we infer that $\lim_{\lambda \to 0^+} T_\lambda x = f_0$ and $f_0 = f$. This completes the proof.

Proposition 6.1.30. The following conclusions hold:

$$\lim_{\lambda \to \lambda_0} T_\lambda x = T_{\lambda_0} x, \qquad \lim_{\lambda \to \lambda_0} R_\lambda x = R_{\lambda_0} x$$

where $\lambda_0 > 0$ and $x \in D(T)$.

Proof. Since

$$\begin{aligned}
&(J(R_\lambda x - x) - J(R_{\lambda_0} x - x), R_\lambda x - R_{\lambda_0} x) \\
&\leq \frac{\lambda - \lambda_0}{\lambda_0}(J(R_{\lambda_0} x - x), R_\lambda x - R_{\lambda_0} x)
\end{aligned}$$

and

$$(\|R_\lambda x - x\| - \|R_{\lambda_0}x - x\|)^2 \leq (J(R_\lambda x - x) - J(R_{\lambda_0}x) - x), R_\lambda x - R_{\lambda_0}x),$$

we have

$$\lim_{\lambda \to 0^+} \|R_\lambda x - x\| = \|R_{\lambda_0}x - x\|,$$

$$\lim_{\lambda \to 0^+} (J(R_{\lambda_0}x - x), R_\lambda x) - x) = \|R_{\lambda_0}x - x\|^2.$$

Thus the locally uniform convexity of E and E^* imply that

$$\lim_{\lambda \to \lambda_0} T_\lambda x = T_{\lambda_0}x, \quad \lim_{\lambda \to \lambda_0} R_\lambda x = R_{\lambda_0}x.$$

This completes the proof.

Proposition 6.1.31. Let $A : D(A) \subseteq E \to 2^{E^*}$ be a multi-valued mapping of class (S_+). Then we have the following:

(1) If $T : D(A) \to 2^{E^*}$ is an upper semicontinuous operator with closed convex values and T maps each bounded subset of $D(A)$ into a relatively compact subset of E^*, then $T + A$ is a multi-valued mapping of class (S_+);

(2) If $M : D(M) \subseteq E \to 2^{E^*}$ is a maximal monotone operator, then $M_\lambda + A$ is a multi-valued mapping of class (S_+);

(3) If $P : E \to 2^{E^*}$ is a pseudomonotone mapping, then $P + A$ is a mapping of class (S_+);

(4) If $P : E \to 2^{E^*}$ is a bounded generalized pseudomonotone mapping, then $P + A$ is a mapping of class (S_+).

The proof of Proposition 6.1.31 is left to the reader as an exercise.

Proposition 6.1.32. If T_1 and T_2 are two bounded mappings of class (S_+), then $\{tT_1 + (1-t)T_2 : t \in [0,1]\}$ is a homotopy of mappings of class (S_+).

Proof. To show that $tT_1 + (1-t)T_2$ satisfies the conditions (1) and (2) of Definition 6.1.12 is trivial.

Now, suppose that $t_j \to t_0$, $x_j \rightharpoonup x_0$ and $f_j^i \in T_i x_j$ such that

$$\limsup_{j \to \infty} (t_j f_j^1 + (1-t_j) f_j^2, x_j - x_0) \leq 0. \tag{6.1.7}$$

Since T_1 and T_2 are mappings of class (S_+), we have

$$\liminf (f_j^1, x_j - x_0) \geq 0, \quad \liminf (f_j^2, x_j - x_0) \geq 0.$$

By virtue of (6.1.7), we get $\lim_{j\to\infty}(f_j^1, x_j - x_0) = 0$ or $\lim_{j\to\infty}(f_j^2, x_j - x_0) = 0$. Hence we have $x_j \to x_0$ and

$$\lim_{j\to\infty}(f_j^1, x_j - x_0) = 0, \quad \lim_{j\to\infty}(f_j^2, x_j - x_0) = 0.$$

Therefore, $\{f_j^1\}$ and $\{f_j^2\}$ have subsequences $\{f_{j_k}^1\}$ and $\{f_{j_k}^2\}$ that converge weakly to $f^1 \in T_1 x_0$ and $f^2 \in T_2 x_0$, respectively. Therefore, we have

$$t_{j_k} f_{j_k}^1 + (1 - t_{j_k}) f_{j_k}^2 \rightharpoonup t_0 f^1 + (1-t) f^2 \in t_0 T_1 x_0 + (1 - t_0) T_2 x_0.$$

This completes the proof.

6.2 Degree Theory for Mappings of Class $(\mathbf{S}_+)$

In this section, we present a degree theory for multi-valued mappings of class (S_+). In the single valued case it was constructed by Browder [35], and for a degree theory for other monotone-type mapping, see [51]. For some notations, we refer to Section 6.1.

Lemma 6.2.1. Let F be a finite dimensional subspace, $\Omega \subset F$ be an open bounded subset and $0 \in \Omega$. Let $T : \overline{\Omega} \to 2^{F^*}$ be an upper semicontinuous mapping with compact convex values, F_0 be a proper subspace of F, $\Omega_{F_0} = \Omega \cap F_0 \neq \phi$ and $T_{F_0} = j_{F_0}^* T : \overline{\Omega_{F_0}} \to 2^{F_0^*}$ be the Galerkin approximation of T, where $j_{F_0}^*$ is the adjoint mapping of natural inclusion $j_{F_0} : F_0 \to F$. If $deg(T, \Omega, 0) \neq deg(T_{F_0}, \Omega_{F_0}, 0)$. Then there exist $x \in \partial\Omega$ and $f \in Tx$ such that

$$(f, x) \leq 0, \quad (f, v) = 0 \quad \text{for all } v \in F_0,$$

where $deg(\cdot,\cdot,\cdot)$ is the topological degree for upper semicontinuous mappings with compact convex values in finite dimensional spaces.

Proof. Since F is finite dimensional, we may assume that F is a Hilbert space and hence $F^* = F$ and $j_{F_0}^*$ is the projection $P : F \to F_0$. Let $F_1 = F \ominus F_0$ and then $F = F_0 \oplus F_1$. Let B_1 be the open unit ball of F_1 and then $\Omega_1 = \Omega_{F_0} \oplus B_1$ is an open subset of F.

Now, we define a mapping $T_1 : \overline{\Omega_1} \to 2^F$ by

$$T_1(u + v) = T_{F_0} u + v \quad \text{for all } u \in \overline{\Omega_{F_0}},\ v \in \overline{B_1}.$$

Then we have

$$deg(T_1, \Omega_1, 0) = deg(T_{F_0}, \Omega_{F_0}, 0) \neq deg(T, \Omega_1, 0).$$

Again, we define a mapping $T^* : \overline{\Omega} \to 2^F$ by

$$T^* u = PTu + (I - P)u \quad \text{for all } u \in \overline{\Omega}.$$

One can easily see that

$$deg(T_1, \Omega_1, 0) = deg(T_1, \Omega \cap \Omega_1, 0),$$

$$deg(T^*, \Omega, 0) = deg(T^*, \Omega \cap \Omega_1, 0).$$

Consider the homotopy class $\{T_t : 0 \leq t \leq 1\}$, where $T_t : \overline{\Omega \cap \Omega_1} \to 2^F$ is defined by $T_t u = tT_1 u + (1-t)T^*$ for all $u \in \overline{\Omega \cap \Omega_1}$ and $0 \leq t \leq 1$. It is easy to see that, if $0 \in T_{t_0} u_0$, then we have

$$u_0 \in F_0 \cap \overline{\Omega \cap \Omega_1}$$

and so

$$\begin{aligned} deg(T^*, \Omega, 0) &= deg(T^*, \Omega \cap \Omega_1, 0) \\ &= deg(T_1, \Omega \cap \Omega_1, 0) \\ &= deg(T, \Omega_1, 0) \\ &= deg(T_0, \Omega_0, 0). \end{aligned}$$

Therefore, we get

$$deg(T^*, \Omega, 0) \neq deg(T, \Omega, 0). \tag{6.2.1}$$

Let

$$H_t u = (tI + (1-t)P)Tu + (1-t)(I-P)u \quad \text{for all } (t, u) \in [0,1] \times \overline{\Omega}.$$

It follows from (6.2.1) that there exist $t_1 \in [0,1]$ and $u_1 \in \partial\Omega$ such that $0 \in H_{t_1} u_1$. Hence there exists $f_1 \in Tu_1$ such that

$$tf_1 + (1-t_1)Pf_1 + (1-t)(I-P)u_1 = 0. \tag{6.2.2}$$

Multiplying (6.2.2) by $v \in F_0$ and u_1, respectively, we obtain

$$(f_1, v) = 0, \quad (f_1, u_1) = 0.$$

This completes the proof.

Lemma 6.2.2. Let $\{T_t\}_{t \in [0,1]}$ be a homotopy of mappings of class (S_+). If $0 \notin T_t(\partial\Omega)$ for all $t \in [0,1]$, then there exists a finite dimensional subspace F_0 such that $0 \notin T_{t,F}(\partial\Omega)$ for all F with $F_0 \subset F$ and $dim F < \infty$, where $T_{t,F} = j_F^* T_t$ for all $t \in [0,1]$.

Proof. Suppose that the conclusion is not true. For finite dimensional subspaces F_0 and F with $F_0 \subset F$, we define a set W_F as follows:

$$\begin{aligned} W_F = \{(t, x) \in [0,1] \times \partial\Omega :\ & \text{there exists } f \in T_t x \\ & \text{such that } (f, x) \leq 0 \text{ and } (f, v) = 0 \text{ for all } v \in F\}. \end{aligned}$$

Then W_F is nonempty. Let $\overline{W_F}$ be the closure of W_F in $[0,1] \times E$ with E endowed with weak topology.

Consider the following family of sets:

$$\mathcal{F} = \{\overline{W_F} : F_0 \subset F, \ dim(F) \leq \infty\}.$$

It is easy to show that $\cap_{F \in \mathcal{F}} \overline{W_F} \neq \phi$. Let $(t_0, x_0) \in \cap_{F \in \mathcal{F}} \overline{W_F}$. For all $v \in E$, if we take a finite dimensional subspace F such that $F_0 \subset F$, $v \in F$ and $x_0 \in F$, then there exist $(t_j^v, x_j^v) \in W_F$ and $f_j^v \in T_{t_j^v} x_j^v$ such that

$$t_j^v \to t_0, \quad x_j^v \rightharpoonup x_0, \quad (f_j^v, x_j^v) \leq 0, \quad (f_j^v, v) = 0$$

for $j = 0, 1, 2, \cdots$. Hence we have $\limsup_{j \to \infty} (f_j^v, x_j^v - x_0) \leq 0$. But since $\{T_t : t \in [0,1]\}$ is a homotopy of mappings of class (S_+), it follows that $x_j^v \to x_0 \in \partial\Omega$ and (f_j^v) has a subsequence $(f_{j_k}^v)$ that converges weakly to $f_0^v \in T_{t_0} x_0$. Therefore, we have $(f_0^v, v) = 0$ for all $v \in E$. By Mazur's separation theorem, we get $0 \in T_{t_0} x_0$, which is a contradiction. This completes the proof.

Lemma 6.2.3. Let $\{T_t\}_{t \in [0,1]}$ be a homotopy of mappings of class (S_+). If $0 \notin T_t(\partial\Omega)$ for all $t \in [0,1]$, then there exists a finite dimensional subspace F_0 such that the topological degree $deg(T_{t,F}, \Omega_F, 0)$ is well defined and does not depend on $t \in [0,1]$ and each finite dimensional subspace F with $F_0 \subset F$.

Proof. By Lemma 6.2.2, there exists a finite dimensional subspace F' such that $0 \notin T_{t,F}(\partial\Omega)$ for all $t \in [0,1]$ and F with $F' \subset F$ and $dimF < \infty$. It is easy to see $T_{t,F} : \overline{\Omega_F} \to 2^{F^*}$ is upper semicontinuous with compact convex values. Thus, by Definition 6.3.12, the topological degree $deg(T_{t,F}, \Omega_F, 0)$ is well defined.

Now, we show that there exists a finite dimensional subspace F_0 such that $F' \subset F_0$, $deg(T_{t,F}, \Omega, 0)$ does not depend on $t \in [0,1]$ and F with $F_0 \subset F$ and $dimF < \infty$. In fact, if this is not true, then, as in the proof of Lemma 6.2.2, we define

$$\begin{aligned} W_F = \{(t,x) \in [0,1] \times \partial\Omega : \ & \text{there exists } f \in T_t x \\ & \text{such that } (f,x) \leq 0 \text{ and } (f,v) = 0 \text{ for all } v \in F\}. \end{aligned}$$

Then W_F is nonempty by Lemma 6.2.1. Let $\overline{W_F}$ be the closure of W_F in $[0,1] \times E$ with E endowed with the weak topology.

Consider again the following family of sets:

$$\mathcal{F} = \{\overline{W_F} : F_0 \subset F, \ dim(F) \leq \infty\}.$$

It is easy to show that $\cap_{F \in \mathcal{F}} \overline{W_F} \neq \phi$. Let $(t_0, x_0) \in \cap_{F \in \mathcal{F}} \overline{W_F}$. Then, for all $v \in E$, if we take a finite dimensional subspace F such that $F_0 \subset F$, $v \in F$ and $x_0 \in F$, then there exist $(t_j^v, x_j^v) \in W_F$ and $f_j^v \in T_{t_j^v} x_j^v$ such that

$$t_j^v \to t_0, \quad x_j^v \rightharpoonup x_0, \quad (f_j^v, x_j^v) \leq 0, \quad (f_j^v, v) = 0$$

for $j = 0, 1, 2, \cdots$. Hence we have $\lim_{j\to\infty}(f_j^v, x_j^v - x_0) \leq 0$. But since $\{T_t : t \in [0,1]\}$ is a homotopy of mappings of class (S_+), we have $x_j^v \to x_0 \in \partial\Omega$ and (f_j^v) has a subsequence $(f_{j_k}^v)$ which converges weakly to $f_0^v \in T_{t_0}x_0$. Therefore, we have $(f_0^v, v) = 0$ for all $v \in E$ and so, by Mazur's separation theorem, $0 \in T_{t_0}x_0$, which is a contradiction. This completes the proof.

Now, let $T : \overline{\Omega} \to 2^{E^*}$ be a mapping of class (S_+) with $0 \notin \partial\Omega$. By Lemma 6.2.3, there exists a finite dimensional subspace F_0 such that $0 \notin T_F(\partial\Omega)$ for all F with $F_0 \subset F$ and $dim F < \infty$, and the topological degree $deg(T_F, \Omega_F, 0)$ is well defined and does not depend on F.

Now, we define the topological degree $deg(T, \Omega, 0)$ as the common value of $deg(T_F, \Omega_F, 0)$.

Theorem 6.2.4. Let E be a reflexive Banach space, $\Omega \subset E$ be an open bounded subset and $T : \overline{\Omega} \to 2^{E^*}$ be a mapping of class (S_+). If $0 \notin T(\partial\Omega)$, then the topological degree $deg(T, \Omega, 0)$ defined as above has the following properties:

(1) $deg(J, \Omega, 0) = 1$ if $0 \in J(\Omega)$;

(2) If $deg(T, \Omega, 0) \neq 0$, then $0 \in Tx$ has a solution in Ω;

(3) If Ω_1 and $\Omega_2 \subset \Omega$ are two open subsets with $\Omega = \Omega_1 \cup \Omega_2$ and $\Omega_1 \cap \Omega_2 = \phi$, then
$$deg(T, \Omega, 0) = deg(T, \Omega_1, 0) + deg(T, \Omega_2, 0);$$

(4) If $\{T_t\}_{t\in[0,1]}$ is a homotopy of mappings of class (S_+), $\Omega \subset D(T_t)$ and $0 \notin T_t(\partial\Omega)$ for all $t \in [0,1]$, then $deg(T_t, \Omega, 0)$ does not depend on $t \in [0,1]$.

6.3 Degree for Perturbations of Monotone-Type Mappings

In this section, based on the results in Sections 6.1 and 6.2, we establish various topological degree theories for monotone type mappings. In the sequel, $T : \overline{\Omega} \to 2^{E^*}$ is a bounded mapping of class (S_+), where Ω is an open bounded subset in E.

Lemma 6.3.1. Let $M : D(M) \subseteq E \to 2^{E^*}$ be a maximal monotone mapping with $D(M) \cap \Omega \neq \emptyset$. If $0 \notin (M + T)(\partial\Omega \cap D(M))$, then there exists $\lambda_0 > 0$ such that
$$0 \notin (M_\lambda + T)(\partial\Omega) \quad \text{for all } \lambda < \lambda_0,$$

where M_λ is the Yosida approximation of M.

Proof. Suppose this is not true. Then there exist $x_j \in \partial\Omega$ with $x_j \rightharpoonup x_0$, $\lambda_j \to 0^+$, $f_j \in Tx_j$ with $f_j \rightharpoonup f_0$ such that

$$M_{\lambda_j}x_j + f_j = 0.$$

We have $(M_{\lambda_j}x_j + f_j, x_j - x_0) = 0$. Since T is a mapping of class (S_+), we have $\liminf_{j\to\infty}(f_j, x_j - x_0) \geq 0$. Thus we get

$$\limsup_{j\to\infty}(M_{\lambda_j}x_j, x_j - x_0) \leq 0.$$

Hence

$$\limsup_{j\to\infty}(M_{\lambda_j}x_j, x_j) \leq (-f_0, x_0). \tag{6.3.1}$$

On the other hand, it follows that $(m - M_{\lambda_j}x_j, x - R_{\lambda_j}x_j) \geq 0$ for all $x \in D(M)$ and $m \in Mx$, i.e.,

$$(m - M_{\lambda_j}x_j, x - x_j + \lambda_j J^{-1}M_{\lambda_j}x_j \geq 0. \tag{6.3.2}$$

By (6.3.1) and (6.3.2), we get

$$(m, x - x_0) - (-f_0, x) \geq -(-f_0, x_0), x \in D(M), m \in Tx.$$

Thus $x_0 \in D(M)$, and $-f_0 \in Mx_0$. Notice that $(M_{\lambda_j}x_j - M_{\lambda_j}x_0, x_j - x_0) \geq 0$. By Lemma 6.1.29, we get

$$\liminf_{j\to\infty}(M_{\lambda_j}x_j, x_j - x_0) \geq 0$$

and so we have

$$\limsup_{j\to\infty}(f_j, x_j - x_0) \leq 0.$$

Therefore, it follows that $x_j \to x_0 \in \partial\Omega \cap D(M)$, $f_0 \in Tx_0$ and $0 \in Mx_0 + Tx_0$, which is a contradiction.

Now, suppose that T and M satisfy the conditions of Lemma 6.3.1. By Lemma 6.3.1, there exists $\lambda_0 > 0$ such that

$$0 \notin (M_\lambda + T)(\partial\Omega \cap D(M)).$$

Since $M_\lambda + T$ is a mapping of class of (S_+), $deg(M_\lambda + T, \Omega, 0)$ is well defined for any $\lambda < \lambda_0$. By Proposition 6.1.30, it is easy to check that

$$\{M_{t\lambda_1 + (1-t)\lambda_2} + T : t \in [0, 1]\}$$

is a homotopy of mappings of class (S_+). Therefore, $deg(M_\lambda + T, \Omega, 0)$ does not depend on $\lambda \in (0, \lambda_0)$.

Now, we define

$$deg(M+T,\Omega\cap D(T),0)=\lim_{\lambda\to 0^+} deg(M_\lambda+T,\Omega,0). \tag{6.3.3}$$

Theorem 6.3.2. The topological degree defined by (6.3.3) has the following properties:

(1) $deg(J,\Omega,0)=1$ if and only if $0\in J(\Omega)$;

(2) If $deg(M+T,\Omega,0)\neq 0$, then $0\in(M+T)x$ has a solution in Ω;

(3) If Ω_1 and $\Omega_2\subset\Omega$ are two open subsets with $\Omega=\Omega_1\cup\Omega_2$ and $\Omega_1\cap\Omega_2=\phi$, then

$$deg(M+T,\Omega,0)=deg(M+T,\Omega_1,0)+deg(M+T,\Omega_2,0);$$

(4) If $\{T_t\}_{t\in[0,1]}$ is a homotopy of mappings of class (S_+), M is a maximal monotone mapping, $\Omega\cap D(M)\neq\emptyset$, $\Omega\subset D(T_t)$ and $0\notin(M+T_t)(\partial\Omega\cap D(M))$ for any $t\in[0,1]$, then

$$deg(M+T_0,\Omega,0)=deg(M+T_1,\Omega,0).$$

Proof. The proof is straightforward.

Remark. The condition (4) of Theorem 6.3.2 is not in its most general form. In applications one may find different homotopy properties in which M may depend on t.

Lemma 6.3.3. Let $M:D(M)\subseteq E\to 2^{E^*}$ be a maximal monotone mapping. If $x_j\in D(T)\cap\overline{\Omega}$, $x_j\rightharpoonup x_0$, (ϵ_j) is a positive sequence with $\epsilon_j\to 0^+$ and $0\in(M+\epsilon_jT)x_j$, then $x_j\to x_0\in D(M)$ and $0\in Mx_0$.

Proof. By our assumptions, there exists $f_j\in Tx_j$ such that

$$(\epsilon_jf_j-\epsilon_if_i,x_j-x_i)\leq 0 \quad \text{for } i,j=1,2,\cdots.$$

By letting $i\to\infty$, we get

$$\epsilon_j(f_j,x_j-x_0)\leq 0.$$

But since T is a mapping of a class (S_+), and so $x_j\to x_0$, we have $x_0\in D(M)$ and $0\in Mx_0$. This completes the proof.

Lemma 6.3.4. Let T_1, T_2 be two mappings of class (S_+), $\epsilon_i>0$ and $\lambda_i>0$ for $i=1,2$. Then $\{M_{t\lambda_1+(1-t)\lambda_2}+t\epsilon_1T_1+(1-t)\epsilon_2T_2:t\in[0,1]\}$ is a homotopy of mappings of class (S_+).

Proof. Let $t_j \to t_0$ and $x_j \rightharpoonup x_0$ be such that

$$\overline{\lim_{j\to\infty}}(M_{t_j\lambda_1+(1-t_j)\lambda_2}x_j + t_j\epsilon_j f_j^1 + (1-t_j)f_j^2, x_j - x_0) \le 0.$$

By the monotonicity of M and Proposition 6.1.30, we have

$$\underline{\lim}_{j\to\infty}(M_{t_j\lambda_1+(1-t_j)\lambda_2}x_j, x_j - x_0) \ge 0.$$

Consequently, we have

$$\overline{\lim_{j\to\infty}}(t_j\epsilon_j f_j^1 + (1-t_j)f_j^2, x_j - x_0) \le 0.$$

But, since $\{t\epsilon_1 T_1 + (1-t)\epsilon_2 T_2 : t \in [0,1]\}$ is a homotopy of mappings of class (S_+), it follows that $x_j \to x_0$, (f_j^1) and (f_j^2) have subsequences $(f_{j_k}^1)$ and $(f_{j_k}^2)$ such that $f_{j_k}^1 \rightharpoonup f^1 \in T_1x_0$ and $f_{j_k}^2 \rightharpoonup f^2 \in T^2x_0$, respectively. Moreover, we have

$$\lim_{j\to\infty}(M_{t_j\lambda_1+(1-t_j)\lambda_2}x_j = M_{t_0\lambda_1+(1-t_0)\lambda_2}x_0.$$

This completes the proof.

Now, suppose that $0 \notin M(\partial\Omega \cap D(M))$. By Lemma 6.3.3, there exists $\epsilon_0 > 0$ such that

$$0 \notin (M + \epsilon T)(\partial\Omega \cap D(M)) \quad \text{for all } \epsilon \in (0, \epsilon_0)$$

and so $deg(M + \epsilon T, \Omega \cap D(M), 0)$ is well defined for any $\epsilon \in (0, \epsilon_0)$. By Lemma 6.3.4, $deg(M + \epsilon T, \Omega \cap D(M), 0)$ does not depend on $\epsilon \in (0, \epsilon_0)$.

Now, we define

$$deg(M, \Omega \cap D(M), 0) = \lim_{\epsilon\to 0} deg(M + \epsilon T, \Omega \cap D(M), 0). \tag{6.3.4}$$

Theorem 6.3.5. The topological degree defined by (6.3.4) has the following properties:

(1) $deg(J, \Omega, 0) = 1$ if and only if $0 \in J(\Omega)$;

(2) If $deg(M, \Omega, 0) \neq 0$, then $0 \in Mx$ has a solution in Ω;

(3) If Ω_1 and $\Omega_2 \subset \Omega$ are two open subsets with $\Omega = \Omega_1 \cup \Omega_2$ and $\Omega_1 \cap \Omega_2 = \emptyset$, then
$$deg(M, \Omega, 0) = deg(M, \Omega_1, 0) + deg(M, \Omega_2, 0);$$

(4) If M_1, M_2 are two maximal monotone mappings, $\Omega \cap D(M_1) \cap D(M_2) \neq \emptyset$ and
$$0 \notin (tM_1 + (1-t)M_2)(\partial\Omega \cap D(M_1) \cap D(M_2)) \cup (tM_{1,\lambda} + (1-t)M_{2,\lambda})(\partial\Omega)$$
for all $t \in [0,1]$ and $\lambda \in (0, \lambda_0)$, then
$$deg(M_1, \Omega, 0) = deg(M_2, \Omega, 0).$$

Proof. The proof follows from (6.3.4), (6.3.3), and Theorem 6.2.4.

Remark. A degree theory can also be developed for pseudomonotone mappings and generalized pseudomonotone mappings and their perturbations with a maximal monotone mappings by using Proposition 6.1.31, which is a method similar to the one employed above, and so we leave it to the reader as an exercise.

6.4 Degree Theory for Mappings of Class $(S_+)_L$

In this section, we construct a topological degree theory for multi-valued mappings of class $(S_+)_{0,L}$, and the topological degree for multi-valued mappings of class $(S_+)_L$ follows from definition 6.1.16.

In the following, we assume that E is also separable, $L \subset E$ is a dense subspace of E and $\Omega \subset E$ is a nonempty open bounded subset. Let $\{T_t\}_{t\in[0,1]}$ be a homotopy of mappings of class $(S_+)_{0,L}$ which has a common domain D and $\Omega_F = \Omega \cap D \cap F$ for each finite dimensional subspace F of L is open in F. We may choose a sequence of finite dimensional subspaces F_j of L such that $F_j \subseteq F_{j+1}$ and $\overline{\cup_{j=1}^{\infty} F_j} = E$. This is possible because E is separable and L is dense in E.

Lemma 6.4.1. Let $T : \overline{\Omega} \cap D(T) \to 2^{E^*}$ be a mapping of $(S_+)_{0,L}$. Suppose that $0 \notin T(\partial\Omega \cap D(T))$. Then there exists an integer $N > 0$ such that

$$0 \notin T_n(\partial\Omega \cap D(T) \cap F_n) \quad \text{for all } n > N,$$

where $T_n = j_{F_n}^* T$.

Proof. Suppose that the conclusion is not true. Then there exists $x_{n_k} \in \partial\Omega \cap D(T) \cap F_{n_k}$ such that $0 \in T_{n_k} x_{n_k}$, i.e., there exists $f_{n_k} \in T x_{n_k}$ such that $0 = j_{F_{n_k}}^* f_{n_k}$ for $k = 1, 2, \cdots$. Without loss of generality, we may assume that $x_{n_k} \rightharpoonup x_0$. Now we have $(f_{n_k}, x) = 0$ for all $x \in F_{n_k}$, $k = 1, 2, \cdots$. Thus $(f_{n_k}, x_{n_k}) = 0$ and $\lim_{k\to\infty}(f_{n_k}, v) = 0$ for all $v \in \cup_{j=1}^{\infty} F_j$. Since T is a mapping of class $(S_+)_{0,L}$, it follows that $x_{n_k} \to x_0 \in \partial\Omega \cap D$ and $0 \in T x_0$, which is a contradiction. This completes the proof.

Under the conditions of Lemma 6.4.1, we know from Section 2.3 that the topological degree $deg(T_n, \Omega \cap D(T) \cap F_n, 0)$ is well defined for sufficiently large n, and we have the following:

Lemma 6.4.2. Let T be the same as in Lemma 6.4.1. Then there exists an integer $N > 0$ such that the topological degree $deg(T_n, \Omega \cap D(T) \cap F_n, 0)$ does not depend on $n > N$, where $T_n = j_{F_n}^* T$.

Proof. Suppose that the conclusion is not true. By Lemma 6.2.1, there exist $x_{n_k} \in \partial\Omega \cap D \cap F_{n_k}$, $f_{n_k} \in Tx_{n_k}$ such that $(f_{n_k}, x_{n_k}) \leq 0$ and $(f_{n_k}, x) = 0$ for all $x \in F_{n_k}$, $k = 1, 2, \cdots$. We may assume that $x_{n_k} \rightharpoonup x_0$. By the same proof as in Lemma 6.4.2, we get $x_{n_k} \to x_0 \in \partial\Omega \cap D$ and $0 \in Tx_0$, which is a contradiction. This completes the proof.

Suppose that $\{E_j\}_{j=1}^{\infty}$ is another sequence of finite dimensional subspaces of L and $\overline{\cup_{j=1}^{\infty} E_j} = E$. Then we have the following:

Lemma 6.4.3. Let T be the same as in Lemma 6.4.1. Then there exists an integer $N > 0$ such that

$$deg(T_n, \Omega \cap D(T) \cap F_n, 0) = deg(T_n, \Omega \cap D(T) \cap E_n, 0) \text{ for all } n > N.$$

Proof. Put $K_n = E_n \cup F_n$. By using the same proof as in Lemma 6.4.2, there exist two integers $N_1 > 0$ and $N_2 > 0$ such that

$$deg(T_n, \Omega \cap D(T) \cap F_n, 0) = deg(T_n, \Omega \cap D(T) \cap K_n, 0) \text{ for all } n > N_1$$

and

$$deg(T_n, \Omega \cap D(T) \cap E_n, 0) = deg(T_n, \Omega \cap D(T) \cap K_n, 0) \text{ for all } n > N_2.$$

Therefore, the conclusion of Lemma 6.4.3 is true. This completes the proof.

Now, let L be a dense subspace of E, $\Omega \subset E$ be a nonempty open bounded subset and $T : D(T) \subset E \to 2^{E^*}$ be a mapping of class $(S_+)_{0,L}$. Assume that $\Omega \cap D(T) \cap F$ is open in F for each finite dimensional subspace F of L. Suppose that $0 \notin T(\partial\Omega \cap D(T))$. In view of lemmas 6.4.1 to 6.4.3, we may define the topological degree by

$$Deg(T, \Omega \cap D(T), 0) = \lim_{n \to \infty} deg(T_n, \Omega \cap D(T) \cap F_n, 0).$$

In general, if T is a mapping of class $(S_+)_L$ and $p \notin T(\partial\Omega \cap D(T))$, then we can define the topological degree by

$$Deg(T, \Omega \cap D(T), p) = Deg(T - p, \Omega \cap D(T), 0).$$

The topological degree defined above has the following properties:

Theorem 6.4.4. Let E be a reflexive Banach space, L be a dense subspace of E, $\Omega \subset E$ be an open bounded subset and $T : D(T) \cap \overline{\Omega} \to 2^{E^*}$ be a mapping of class $(S_+)_{0,L}$. If $0 \notin T(\partial\Omega \cap D(T))$, then the topological degree $Deg(T, \Omega \cap D(T), 0)$ defined above has the following properties:

(1) $Deg(J, \Omega, 0) = 1$ if $0 \in J(\Omega)$;

(2) If $Deg(T, \Omega \cap D(T), 0) \neq 0$, then $0 \in Tx$ has a solution in $\Omega \cap D(T)$;

(3) If Ω_1 and $\Omega_2 \subset \Omega$ are two open subsets with $\Omega = \Omega_1 \cup \Omega_2$ and $\Omega_1 \cap \Omega_2 = \emptyset$, then

$$Deg(T, \Omega \cap D(T), 0) = Deg(T, \Omega_1 \cap D(T), 0) + Deg(T, \Omega_2 \cap D(T), 0);$$

(4) If $\{T_t\}_{t\in[0,1]}$ is a homotopy of mappings of class $(S_+)_{0,L}$ with $D(T_t) = D$ and $0 \notin T_t(\partial\Omega \cap D)$ for all $t \in [0, 1]$, then $Deg(T_t, \Omega \cap D, 0)$ does not depend on $t \in [0, 1]$.

Proof. (1) to (3) follow easily from the definition and the properties of degree theory in finite dimensional spaces.

We only need to prove (4). Assume that $(F_j)_{j=1}^{\infty}$ is a sequence of finite dimensional subspaces of L with $\overline{\cup_{j=1}^{\infty} F_j} = E$. We claim that there exists an integer $N > 0$ such that

$$0 \notin T_{t,n}(\partial\Omega \cap D \cap F_n) \quad \text{for all } n > N,\ t \in [0, 1],$$

where $T_{t,n} = j_{F_n}^* T_t$. If not, then there exist $t_{n_k} \to t_0$, $x_{n_k} \in \partial\Omega \cap D \cap F_{n_k}$ with $x_{n_k} \rightharpoonup x_0$, $f_{n_k} \in T_{t_{n_k}} x_{n_k}$ such that $0 = j_{F_{n_k}}^* f_{n_k}$, which implies that

$$(f_{n_k}, x_{n_k}) = 0, \quad (f_{n_k}, v) = 0 \text{ for all } v \in F_{n_k}.$$

Since $\{T_t\}_{t\in[0,1]}$ is a homotopy of mappings of class $(S_+)_{0,L}$, we get $x_{n_k} \to x_0 \in \partial\Omega \cap D$ and $0 \in T_{t_0} x_0$, which is a contradiction. Thus the claim is true.

Now, for all $n > N$, we know from Section 2.3 that $deg(T_{t,n}, \Omega \cap D \cap L \cap F_n, 0)$ is a constant for $t \in [0, 1]$, where $T_{t,n} = j_{F_n}^* T_t$. In view of Lemma 6.4.3, we see that the conclusion of (4) is true. This completes the proof.

Theorem 6.4.5. Let $T : D(T) \subset E \to 2^{E^*}$ be a mapping of class $(S_+)_{0,L}$ and $\Omega \subset E$ be an open bounded subset such that $\Omega \cap D(T) \cap F$ is open in F for each finite dimensional subspace F of L. If $0 \in \Omega \cap D(T)$ and $(f, x) > 0$ for all $x \in \partial\Omega \cap D(T)$ and $f \in Tx$, then

$$Deg(T, \Omega \cap D(T), 0) = 1.$$

Proof. Assume that $(F_j)_{j=1}^{\infty}$ is a sequence of finite dimensional subspaces of L with $\overline{\cup_{j=1}^{\infty} F_j} = E$. It is easy to check that

$$0 \notin (t j_{F_n}^* T + (1 - t) j_{F_n}^* J)(\partial\Omega \cap D(T) \cap F_n)$$

for all $t \in [0, 1]$ and $n = 1, 2, \cdots$. Therefore, we have

$$deg(j_{F_n}^* T, \Omega \cap D(T) \cap F_n, 0) = deg(j_{F_n}^* J, \Omega \cap D(T) \cap F_n, 0) = 1$$

for all $n = 1, 2, \cdots$ such that $F_n \cap \Omega \cap D(T) \neq \emptyset$. By the definition of topological degree $Deg(T, \Omega \cap D(T), 0)$, we get

$$Deg(T, \Omega \cap D(T), 0) = 1.$$

This completes the proof.

Theorem 6.4.6. Let H be a real separable Hilbert space, L be a dense subspace of H, $\Omega \subset H$ be an open bounded subset such that $\Omega \cap D(T) \cap F$ is open in F for each finite dimensional subpace F of L and $T : D(T) \to 2^H$ be a L_0-pseudocompact mapping. Suppose that $0 \in D(T) \cap \Omega$ and

$$(Tx, x) \leq \|x\|^2 \quad \text{for all } x \in \partial\Omega \cap D(T).$$

Then T has a fixed point in $\overline{\Omega} \cap D(T)$.

Proof. We may assume that $x \notin Tx$ for all $x \in \partial\Omega \cap D(T)$. Assume that $(F_j)_{j=1}^{\infty}$ is a sequence of finite dimensional subspaces of L with $\overline{\cup_{j=1}^{\infty} F_j} = H$. We claim that there exists an integer $N > 0$ such that

$$0 \notin [(tP_n(I - T) + (1 - t)P_n](\partial\Omega \cap D(T) \cap F_n$$

for all $n > N$ and $t \in [0, 1]$, where $P_n : H \to F_n$ is the projection. Assume this is not true. Then there exist $t_j \to t_0$, $x_j \in \partial\Omega \cap D \cap F_{n_j}$ with $x_j \rightharpoonup x_0$ and $f_j \in Tx_j$ such that $P_{F_{n_j}}(x_j - t_j f_j) = 0$ for $j = 1, 2, \cdots$, where $P_{F_{n_j}} : H \to F_{n_j}$ is the projection. Thus it follows that $(x_j - t_j f_j, x_j) = 0$. This and our assumption imply that $t_j = 1$, so we have

$$(x_j - f_j, x_j) = 0, \quad (x_j - f_j, v) = 0 \text{ for all } v \in F_{n_j},\ j = 1, 2, \cdots .$$

From which we get $x_j \to x_0 \in \partial\Omega \cap D(T)$ and $x_0 \in Tx_0$, which is a contradiction. Thus we get

$$deg(P_n(I - T), \Omega \cap D(T) \cap F_n, 0) = deg(P_n, \Omega \cap F_n, 0) = 1.$$

Thus it follows that $Deg(I - T, \Omega \cap D(T), 0) = 1$. By Theorem 6.4.4, T has a fixed point in $\overline{\Omega} \cap D(T)$. This completes the proof.

6.5 Coincidence Degree for Mappings of Class L-(S_+)

In this section, we construct a topological degree theory for multi-valued mappings of class L-(S_+).

Lemma 6.5.1. Let L be a Fredholm mapping of index zero type, $\Omega \subset H$ be an open bounded subset and $T : \overline{\Omega} \to 2^H$ be a mapping. If $0 \notin (L - T)(\partial\Omega \cap D(L))$, then $0 \notin [I - P - (JQ + K_{PQ})T](\partial\Omega)$.

Proof. Suppose the contrary. Then there exists $x_0 \in \partial\Omega$ such that $0 \in x_0 - Px_0 - (JQ + K_{PQ})Tx_0$, and so there exist $f_0 \in Tx_0$ such that

$$x_0 - Px_0 - JQf_0 - K_{PQ}f_0 = 0.$$

Since

$$JQf_0 \in Ker(L) = Im(P),\ x_0 - Px_0 \in Ker(P),\ K_{pQ}f_0 \in D(L) \cap Ker(P),$$

we must have

$$JQf_0 = 0, x_0 - Px_0 - K_{PQ}f_0 = 0.$$

Therefore, we have

$$Qf_0 = 0, x_0 - Px_0 - K_P f_0 = 0, \quad \text{i.e.,} \quad Lx_0 - f_0 = 0,$$

which is a contradiction to $0 \notin (L-T)(\partial\Omega \cap D(L))$. This completes the proof.

Now, let L be a Fredholm mapping of index zero type, $\Omega \subset H$ be an open bounded subset and $T : \overline{\Omega} \to 2^H$ be a mapping of class L-(S_+). Suppose that $0 \notin (L - T)(\partial\Omega \cap D(L))$. By Lemma 6.5.1, we have $0 \notin [I - P - (JQ + K_{PQ})T](\partial\Omega)$. As a result, $deg(I - P - (JQ + K_{PQ})T, \Omega, 0)$ is well defined. We define

$$deg_J(L - T, \Omega, 0) = deg(I - P - (JQ + K_{PQ})T, \Omega, 0),$$

which is called the coincidence degree of L and T on Ω.

Notice that the projections P, Q in Hilbert spaces are uniquely determined, so $deg_J(L - T, \Omega, 0)$ is well defined.

The following result follows directly from Theorem 6.2.4:

Theorem 6.5.2. The coincidence degree of L and T on Ω has the following properties:

(1) If Ω_1 and Ω_2 are disjoint open subsets of Ω such that $0 \notin (L-T)(D(L) \cap \overline{\Omega \setminus (\Omega_1 \cup \Omega_2)})$, then

$$deg_J(L - T, \Omega, 0) = deg_J(L - T, \Omega_1) + deg_J(L - T, \Omega_2, 0);$$

(2) If $H(t, x) : [0, 1] \times \overline{\Omega} \to Y$ is a mapping of class L-(S_+) on $[0, 1] \times \overline{\Omega}$ and, if $0 \neq Lx - H(t, x)$ for all $(t, x) \in [0, 1] \times \partial\Omega$, then $deg_J(L - H(t, \cdot), \Omega, 0)$ does not depend on $t \in [0, 1]$;

(3) If $deg_J(L - T, \Omega, 0) \neq 0$, then $0 \in (L - T)(D(L) \cap \Omega)$;

(4) If $L : D(L) \subseteq H \to H$ is a linear mapping such that $L^{-1} : H \to D(L)$ is continuous, then $deg_J(L, \Omega, 0) = 1$ if $0 \in \Omega$.

Theorem 6.5.3. Let L be a Fredholm mapping of index zero type with $Ker(L) = \{0\}$, $0 \in \Omega \subset H$ be an open bounded subset and $T : \overline{\Omega} \to 2^H$ be a mapping of class L-(S_+). Suppose that $Lx \notin tTx$ for all $x \in \partial\Omega \cap D(L)$ and $t \in [0, 1]$. Then $deg_J(L - T, \Omega \cap D(T), 0) = 1$.

Proof. Since $Ker(L) = \{0\}$, we take $P = Q = 0$ and $J = I$. Now, $I - L^{-1}T$ is mapping of class (S_+) and $\{I - tL^{-1}T]\}_{t\in[0,1]}$ is a homotopy of mappings of class (S_+). From our assumption, one can easily see that $0 \notin (x - tL^{-1}Tx)$ for all $x \in \partial\Omega$, $t \in [0,1]$. From Theorem 6.2.4, we have

$$deg(I - L^{-1}T, \Omega, 0) = deg(I, \Omega, 0) = 1,$$

i.e., $deg_J(L - T, \Omega \cap D(T), 0) = 1$. This completes the proof.

Corollary 6.5.4. Let L be a Fredholm mapping of index zero type with $Ker(L) = \{0\}$ and $T : H \to 2^H$ be a mapping of class L-(S_+). Then one of the following conditions hold:

(1) $Lx \in Tx$ has a solution in $D(T)$;

(2) $\{x : Lx \in \lambda Tx \text{ for some } \lambda \in (0,1)\}$ is unbounded.

Theorem 6.5.5. Let L be a Fredholm mapping of index zero type, $\Omega \subset H$ an open bounded subset and $T_i : \overline{\Omega} \to 2^H$ be a mapping of class L-(S_+) for $i = 1, 2$. Suppose that $T_1x = T_2x$ for all $x \in \partial\Omega \cap D(L)$. Then we have the following:

$$deg_J(L - T_1, \Omega, 0) = deg_J(L - T_2, \Omega, 0).$$

Proof. Since $T_1x = T_2x$ for all $x \in \partial\Omega$, we have

$$0 \notin (I - P - (JQ + K_{PQ}(tT_1 + (1-t)T_2)))x \quad \text{for all } x \in \partial\Omega.$$

Moreover, J is linear, so

$$\begin{aligned} &I - P - (JQ + K_{PQ})(tT_1 + (1-t)T_2) \\ &= I - P - t(JQ + K_{PQ})T_1 - (1-t)(JQ + K_{PQ})T_2 \end{aligned}$$

is a homotopy of mappings of class (S_+). Therefore, it follows from Theorem 6.2.4 that

$$deg(I - P - (JQ + K_{PQ})T_1, \Omega, 0) = deg(I - P - (JQ + K_{PQ})T_2, \Omega, 0)$$

and so

$$deg_J(L - T_1, \Omega, 0) = deg_J(L - T_2, \Omega, 0).$$

This completes the proof.

Suppose that L is a Fredholm mapping of index zero type, $\Omega \subset H$ be an open bounded subset, $\overline{\Omega} \cap D(L) \cap D(T) \neq \emptyset$ and $T : D(T) \to 2^H$ is a L-maximal monotone mapping. Also, suppose that $0 \notin (L - T)(\partial\Omega \cap D(L) \cap D(T))$. Then, by Lemma 6.5.1, we have

$$0 \notin [I - P - (JQ + K_{PQ})T](\partial\Omega \cap D(T)).$$

Since $I - P - (JQ + K_{PQ})T$ is maximal monotone, the degree $deg(I - P - (JQ + K_{PQ})T, \Omega, 0)$ is well defined, and we may define a degree

$$deg_J(L - T, \Omega, 0) = deg(I - P - (JQ + K_{PQ})T, \Omega, 0),$$

and one may deduce results similar to Theorem 6.5.3.

Lemma 6.5.6. Let $T : D(T) \subseteq H \to 2^H$ be a bounded multi-valued mapping and $0 \notin \overline{(L - T)(\partial\Omega \cap D(L) \cap D(T))}$. Then there exists $\epsilon_0 > 0$ such that

$$0 \notin [I - P - (JQ + K_{PQ})T + \epsilon I](\partial\Omega \cap D(T)) \text{ for all } \epsilon \in (0, \epsilon_0).$$

Proof. Assume that the conclusion is false. Then there exist $\epsilon_n \to 0$ and $x_n \in \partial\Omega \cap D(T)$ such that

$$0 \in (I - P - (JQ + K_{PQ})T + \epsilon_n)x_n.$$

Thus there exist $f_n \in Tx_n$ such that $0 = x_n - Px_n - (JQ + K_{PQ})f_n + \epsilon_n x_n$. Therefore, we have

$$(1 + \epsilon_n)(I - P)x_n + (\epsilon_n Px_n + JQf_n) + K_{PQ}f_n = 0.$$

Consequently, it follows that

$$JQf_n = -\epsilon_n Px_n, \quad (1 + \epsilon_n)(I - P)x_n + K_{PQ}f_n = 0$$

and so we have $(1 + \epsilon_n)Lx_n - f_n - \epsilon_n J^{-1}Px_n = 0$. By the boundedness of T, we get $Lx_n - f_n \to 0$, which is a contradiction. This completes the proof.

Now, we assume that $T : \overline{\Omega} \to H$ is a bounded mapping and T is L-pseudomonotone. Suppose that $0 \notin \overline{(L - T)(\partial\Omega \cap D(L))}$. Then by Lemma 6.5.6, there exists $\epsilon_0 > 0$ such that

$$0 \notin [I - P - (JQ + K_{PQ})T + \epsilon I](\partial\Omega) \quad \text{for all } \epsilon \in [0, \epsilon_0).$$

Since $I - P - (JQ + K_{PQ})T$ is pseudomonotone, $I - P - (JQ + K_{PQ})T + \epsilon I$ is a mapping of class (S_+) and the degree $deg(I - P - (JQ + K_{PQ})T + \epsilon I, \Omega, 0)$ is well defined for all $\epsilon \in (0, \epsilon_0)$. We define

$$deg_J(L - T, \Omega, 0) = \lim_{\epsilon \to 0^+} deg(I - P - (JQ + K_{PQ})T + \epsilon I, \Omega, 0),$$

which is easily seen to be well defined if we note that $\{I - P - (JQ + K_{PQ})T + [t\epsilon_1 - (1 - t)\epsilon_2]I\}_{t \in [0,1]}$ is a homotopy of mappings of class (S_+).

Theorem 6.5.7. The coincidence degree of L and the pseudomonotone mapping T on Ω has the following properties:

(1) If Ω_1 and Ω_2 are open subsets of Ω such that

$$\Omega_1 \cap \Omega_2 = \emptyset, \quad 0 \notin \overline{(L-T)(D(L) \cap \overline{\Omega \setminus (\Omega_1 \cup \Omega_2)}},$$

then we have

$$deg_J(L-T, \Omega, 0) = deg_J(L-T, \Omega_1) + deg_J(L-T, \Omega_2, 0);$$

(2) If $T_i : \overline{\Omega} \to H$, $i = 1, 2$, are bounded mapping such that $tT_1 + (1-t)T_2$ is L-pseudomonotone on $[0,1] \times \overline{\Omega}$ and

$$0 \notin \overline{\cup_{t\in[0,1]}[L - tT_1 + (1-t)T_2](\partial\Omega \cap D(L))},$$

then we have

$$deg_J(L - T_1, \Omega, 0) = deg_J(L - T_2, \Omega, 0);$$

(3) If $deg_J(L-T, \Omega, 0) \neq 0$, then $0 \in \overline{(L-T)(D(L) \cap \Omega)}$.

6.6 Computation of Topological Degree

In this section, we compute the topological degree of some monotone-type mappings under certain boundary conditions.

Theorem 6.6.1. Let $T : \overline{\Omega} \subset E \to 2^{E^*}$ be a mapping of class (S_+). If $0 \in \Omega$ and $(f, x) > 0$ for all $x \in \partial\Omega$ and $f \in Tx$, then $deg(T, \Omega, 0) = 1$.

Proof. Consider the family $\{tT + (1-t)J : t \in [0,1]\}$ of mappings. By our assumptions, we know that $0 \notin (tT + (1-t)J)(\partial\Omega)$ for all $t \in [0,1]$, and so a direct proof shows that $\{tT + (1-t)J : t \in [0,1]\}$ is a homotopy of mappings of class (S_+). By Theorem 6.2.4, we have

$$deg(T, \Omega, 0) = d(J, \Omega, 0) = 1.$$

This completes the proof.

Theorem 6.6.2. Let $T : \overline{\Omega} \subset E \to 2^{E^*}$ be a mapping of class (S_+) and $M : D(M) \subseteq E \to 2^{E^*}$ be a maximal monotone mapping. If $x_0 \in \Omega \cap D(M)$ and

$$(f, x - x_0) > -\|f\|\|x - x_0\|$$

for all $x \in \partial\Omega \cap D(M)$ and $f \in (M+T)x$, then $deg(M+T, \Omega, 0) = 1$.

Proof. It is easy to see that

$$0 \notin (tM + (1-t)T + (1-t)J(\cdot - x_0))(\partial\Omega \cap D(M)) \quad \text{for all } t \in [0,1].$$

Now, we prove that there exists $\lambda_0 > 0$ such that

$$0 \notin (tM_\lambda + (1-t)T + (1-t)J(\cdot - x_0))(\partial\Omega) \text{ for all } t \in [0,1].$$

In fact, if this is not true, then there exist $t_j \to t_0$, $x_j \in \partial\Omega$, $x_j \rightharpoonup y_0$, $f_j \in Tx_j$ and $\lambda_j \to 0^+$ such that

$$0 = t_j M_{\lambda_j} x_j + (1-t_j) f_j + (1-t_j) J(x_j - x_0). \tag{6.6.1}$$

Multiplying (6.6.1) by $x_j - y_0$, we obtain

$$(t_j M_{\lambda_j} x_j + (1-t_j) f_j + (1-t_j) J(x_j - x_0), x_j - y_0) = 0. \tag{6.6.2}$$

We consider the following two cases:
Case (i) $t_0 = 0$. Multiplying (6.6.1) by $x_j - x_0$, we obtain

$$(t_j M_{\lambda_j} x_j + (1-t_j) f_j + (1-t_j) J(x_j - x_0), x_j - x_0) = 0.$$

By Lemma 6.1.29, we have $t_j M_{\lambda_j} x_0 \to 0$ and so we obtain

$$\limsup_{j\to\infty} ((1-t_j) f_j + (1-t_j) J(x_j - x_0), x_j - x_0) = 0.$$

It follows that $x_j \to x_0$, which is a contradiction.

Case (ii) $t_0 \neq 0$. Since T and $J(\cdot - x_0)$ are mappings of class (S_+), it follows that

$$\liminf_{j\to\infty} ((1-t_j) f_j + (1-t_j) J(x_j - x_0), x_j - y_0) \geq 0.$$

By (6.6.2), we have

$$\limsup_{j\to\infty} (t_j M_{\lambda_j} x_j, x_j - y_0) \leq 0. \tag{6.6.3}$$

Without loss of generality, we may assume that $M_{\lambda_j} x_j \rightharpoonup m_0$. By (6.6.3), we have

$$\limsup_{j\to\infty} (M_{\lambda_j} x_j, x_j \leq (m_0, y_0). \tag{6.6.4}$$

The monotonicity of M implies that

$$(M_{\lambda_j} x_j - m, x_j - \lambda_j J^{-1} M_{\lambda_j} x_j - x) \geq 0 \tag{6.6.5}$$

for all $x \in D(M)$ and $m \in Mx$. By letting $j \to \infty$ in (6.6.5), we get

$$\limsup_{j\to\infty} (M_{\lambda_j} x_j, x_j) \geq (m, y_0 - x) + (m_0, x) \tag{6.6.6}$$

for all $x \in D(M)$ and $m \in Mx$. Thus, from (6.6.4) and (6.6.6), it follows that

$$(m_0 - m, y_0 - x) \geq 0 \quad \text{for all } x \in D(M),\ m \in Mx,$$

which implies that $y_0 \in D(M)$. Therefore, (6.6.3) becomes the following:

$$\lim_{j\to\infty}(t_j M_{\lambda_j} x_j, x_j - y_0) = 0$$

and hence we have

$$\lim_{j\to\infty}((1-t_j)f_j + (1-t_j)J(x_j - x_0), x_j - y_0) = 0.$$

Therefore, $x_j \to y_0$, and (6.6.1) implies that

$$0 \in (t_0 M + (1-t_0)T + (1-t_0)J(\cdot - x_0))(\partial\Omega \cap D(M),$$

which is a contradiction. By Theorem 6.2.4, we conclude that

$$deg(M_\lambda + T, \Omega, 0) = deg(J(\cdot - x_0), \Omega, 0) = 1 \quad \text{for all } \lambda \in (0, \lambda_0).$$

By the definition of $deg(T+M, \Omega, 0)$, we finally obtain

$$deg(T+M, \Omega, 0) = 1.$$

This completes the proof.

Theorem 6.6.3. Let $f : D(f) \subseteq E \to R \cup \{+\infty, -\infty\}$ be a proper lower semicontinuous convex function. If $\liminf_{\|x\|\to\infty} f(x) = +\infty$, then

$$\lim_{R\to\infty} deg(\partial f, B(0,R) \cap D(\partial f), 0) = 1,$$

where $B(0,R)$ is the open ball with radius R in E.

Proof. Since f is a lower semicontinuous convex function, ∂f is maximal monotone. By our assumptions, we know that there exists $x_0 \in D(\partial f)$ such that $0 \in \partial f(x_0)$. Hence we have

$$(g, x - x_0) \geq 0 \quad \text{for any } g \in \partial f(x),\ x \in D(\partial f).$$

If we take R sufficiently large such that $0 \notin (D(\partial f) \cap B(0,R))$, then, in addition, we have

$$(g + \epsilon J(x - x_0), x - x_0) > 0 \text{ for all } x \in D(\partial f) \cap B(0,R),\ g \in \partial f.$$

By Theorem 6.5.2, we conclude that

$$deg(\partial f + \epsilon J(\cdot - x_0), B(0,R) \cap D(\partial f), 0) = 1.$$

By the definition of $deg(\partial f, B(0,R) \cap D(\partial f), 0)$, we get

$$deg(\partial f, B(0,R) \cap D(\partial f), 0) = 1.$$

This completes the proof.

Theorem 6.6.4. Let H be real Hilbert space, $0 \in \Omega \subset H$ be an open bounded subset and $T : \overline{\Omega} \to 2^H$ be a pseudocompact mapping. Suppose that

$$\lambda x \notin Tx \quad \text{for all } x \in \partial\Omega,\ \lambda > 1.$$

Then T has a fixed point in $\overline{\Omega}$.

Proof. We may assume that $x \notin Tx$ for all $x \in \partial\Omega$ and $\lambda > 1$. Put $H(t,x) = t(I-T)x + (1-t)x$ for $(t,x) \in [0,1] \times \overline{\Omega}$. Then $H(t,\cdot)$ is a homotopy mapping of a class (S_+). Moreover, we have $0 \notin H(t,x)$ for all $(t,x) \in [0,1] \times \overline{\Omega}$. Therefore, we have

$$deg(I-T,\Omega,0) = deg(I,\Omega,0) = 1.$$

Hence T has a fixed point in $\overline{\Omega}$. This completes the proof.

Theorem 6.6.5. Let H be real Hilbert space, $0 \in \Omega \subset H$ be an open bounded subset and $T : \overline{\Omega} \to 2^H$ be a pseudo-contractive mapping. Suppose that

$$\lambda x \notin Tx \quad \text{for all } x \in \partial\Omega,\ \lambda > 1.$$

Then T has a fixed point in $\overline{\Omega}$.

Proof. We may assume that $x \notin Tx$ for all $x \in \partial\Omega$ and $\lambda > 1$. Put $H(t,x) = t(I-T)x + (1-t)x$ for all $(t,x) \in [0,1] \times \overline{\Omega}$. Then we can obtain the conclusion by using the same argument as in Theorem 6.6.4. This completes the proof.

6.7 Applications to PDEs and Evolution Equations

In this section, we give some applications of our degree theory in the previous sections to periodic nonlinear evolution equations and existence problems of partial differential equations.

In the following, let E be a separable reflexive Banach space which is densely embedded in a real Hilbert space H and E^* be the dual space of E. Let $\|\cdot\|$ and $\|\cdot\|_*$ be the norms in E and E^*, respectively. We always assume that both E and E^* are locally uniformly convex. The following function spaces will be used in the sequel:

$$L^p(0,T;E) = \Big\{ f(t) : [0,T] \to E,\ \int_0^T \|f(s)\|^p ds < \infty \Big\}$$

with the norm $\|f(\cdot)\|_p = (\int_0^T \|f(s)\|^p ds)^{\frac{1}{p}}$ and $L^q(0,T;E^*)$ is the dual space of $L^p(0,T;E)$ with the norm $\|g(\cdot)\|_{*,q} = (\int_0^T \|g(s)\|^q ds)^{\frac{1}{q}}$, where $\frac{1}{p} + \frac{1}{q} = 1$.

For all $u(\cdot) \in L^p(0,T;E)$ and $v(\cdot) \in L^q(0,T;E^*)$, let

$$\langle u, v\rangle = \int_0^T (u(s), v(s))ds$$

and denote

$$W^{1,p}(0,T;E) = \{u(\cdot) \in L^p(0,T;E) : u'(\cdot) \in L^q(0,T;E^*),\ u(0) = u(T)\}$$

with the norm $\|u\|_w = \|u\|_p + \|u'\|_{*,q}$, where u' is the generalized derivative of u and $\|\cdot\|_p$, $\|\cdot\|_{*,q}$ are norms of the spaces $L^p(0,T;E)$ and $L^q(0,T;E^*)$, respectively. Let W^* be the dual space of $W^{1,p}(0,T;E)$.

Theorem 6.7.1. Let $A(t) : E \to 2^{E^*}$ be an operator of class (S_+) for all $t \in R$. Suppose the following conditions are satisfied:

(1) For all $u(t) \in W^{1,p}(0,T;E)$, $A(t)u(t)$ is E^*-measurable on $[0,T]$;

(2) There exist a constant $C > 0$ and $C_1(\cdot) \in L^q(0,T)$ such that

$$\|f\|_* \le C\|x\|^{p-1} + C_1(t)$$

for all $x \in E$ and $f(t) \in A(t)x$ for almost everywhere $t \in [0,T]$;

(3) There exist a constant $\alpha > 0$ and a function $\gamma(\cdot) \in L(0,T)$ such that

$$(f, x) \ge \alpha\|x\|^p - \gamma(t)$$

for all $x \in E$ and $f(t) \in A(t)x$ for almost everywhere $t \in [0,T]$.

Then we have the following:

$$\begin{cases} x'(t) \in -A(t)x(t), & a.e.\, t \in (0,T), \\ x(0) = x(T) \end{cases} \tag{E 6.7.1}$$

has a solution $x(t) \in W^{1,p}(0,T;E)$.

First, we define a multi-valued operator $\mathcal{A} : W^{1,p}(0,T;E) \to W^*$ by

$$\mathcal{A}u(\cdot) = \{f(\cdot) \in W^* : f(t) \in A(t)u(t)\} \text{ for } a.e.\, t \in [0,T]\}.$$

By the condition (1) and Theorem 2.3.9, $\mathcal{A}$ is well defined.

Next, for each $n \ge 1$, we define a multi-valued operator $\mathcal{S}_n : W^{1,p}(0,T;E) \to W^*$ as follows:

For all $u \in W^{1,p}(0,T;E)$ and $g \in \mathcal{S}_n u$, there exists $f \in \mathcal{A}u$ such that

$$(g, v) = (u'(\cdot), v(\cdot)) + (f, v(\cdot)) + \frac{1}{n}(Ju, v) \quad \text{for all } v \in W^{1,p}(0,T;E), \tag{6.7.1}$$

where $J : W^{1,p}(0,T;E) \to W^*$ is the duality mapping.

Lemma 6.7.2. $\mathcal{S}_n$ defined by (6.7.1) is an operator of class (S_+) for each positive integer n.

Proof. The bounded closed convexity of $\mathcal{S}_n u$ is obvious. To prove the finite dimensional weak upper semicontinuity of $\mathcal{S}_n$, we show that $\mathcal{S}_n$ is actually upper semicontinous from the strong topology in $W^{1,p}(0,T;E)$ to the weak topology in W^*. Note that the condition (2) implies that $\mathcal{S}_n$ is bounded and so it is relatively weakly compact in W^*. Hence, we only need to show that $\mathcal{S}_n$ is weakly closed (see [12]). Let $u_n \to u_0$ in $W^{1,p}(0,T;E)$, $g_n \in \mathcal{S}u_n$ and $g_n \rightharpoonup g_0$ in W^*.

Now, we prove that $g_0 \in \mathcal{S}_n u_0$. In fact, there exists $f_n \in \mathcal{A}u_n$ such that

$$f_n \rightharpoonup g_0 - u_0' - \frac{1}{n}Ju_0$$

and there exists a subsequence (u_{n_k}) such that $u_{n_k}(t) \to u_0(t)$ for almost everywhere $t \in [0,T]$. Therefore, it follows from the condition (2) that

$$\lim_{n_k\to\infty} (f_{n_k}(t), u_{n_k}(t) - u_0(t)) = 0$$

for almost everywhere $t \in [0,T]$ and $A(t)$ is a mapping of class (S_+) and so $(f_{n_k}(t))$ has a subsequence converging weakly to $f_0(t) \in A(t)u_0(t)$ for almost everywhere $t \in [0,T]$. Hence we have $g_0 \in \mathcal{S}_n u_0$.

Let $u_j \in W^{1,p}(0,T;E)$ with $u_j \rightharpoonup u_0$ in $W^{1,p}(0,T;E)$ and $g_j \in \mathcal{S}_n u_j$ such that

$$\overline{\lim}_{j\to\infty}(g_j, u_j - u_0) \le 0.$$

Then, by (6.7.1), there exist $f_j \in \mathcal{A}u_j$ such that

$$\overline{\lim}_{j\to\infty} \int_0^T \Big[(f_j, u_j - u_0) + \frac{1}{n}(Ju_j, u_j - u_0)\Big] dt \le 0. \tag{6.7.2}$$

Now, using the fact that $A(t)$ is a mapping of class (S_+), the conditions (2) and (3), we have

$$\underline{\lim}_{j\to\infty}(f_j(t), u_j(t) - u_0(t)) \ge 0 \tag{6.7.3}$$

for almost everywhere $t \in [0,T]$ and thus $u_j \to u_0$ in $W^{1,p}$. There is a subsequence (u_{j_k}) such that $u_{j_k}(t) \to u_0(t)$ for almost everywhere $t \in [0,T]$. Thus, condition (2) implies that

$$\lim_{j\to\infty} (f_{j_k}(t), u_{j_k}(t) - u_0(t)) = 0$$

for almost everywhere $t \in [0,T]$ and $A(t)$ is a mapping of class (S_+) and so (f_{j_k}) has a subsequence $(f_{j_{k'}})$ such that $f_{j_{k'}}(t) \rightharpoonup f_0(t) \in A(t)u_0(t)$. Hence, $(g_{j_{k'}})$ converges weakly to $g_0 \in \mathcal{S}_n u_0$, where g_0 satisfies

$$(g_0, v) = (u_0', v) + (f_0, v) + \frac{1}{n}(Ju_0, v) \quad \text{for all } v \in W^{1,p}(0,T;E).$$

This completes the proof.

Proof of Theorem 6.7.1. First, by the condition (3), we have

$$(g,u) \geq \alpha \int_0^T \|u\|^p dt - \int_0^T \gamma(t)dt + \frac{1}{n}\int_0^T [\|u\|^p + \|u'\|_*^q]dt$$

for all $g \in \mathcal{S}_n u$ and so there exists $r_0^n > 0$ such that

$$(g,u) > 0 \quad \text{for all } u \in W^{1,p}(0,T;E),\ \|u\|_w = r_0^n,\ g \in \mathcal{S}_n u.$$

It is easy to prove that $t\mathcal{S}_n + (1-t)J$ is a homotopy of operators of class (S_+) for all $t \in [0,1]$. By Theorem 6.2.4, we get

$$deg(\mathcal{S}_n, B(0,r_0^n), 0) = deg(J, B(0,r_0^n), 0) = 1,$$

where $B(0,r_0^n)$ is the open ball with radius r_0^n in $W^{1,p}(0,T;E)$. Therefore, $0 \in \mathcal{S}_n u$ has a solution $u_n \in B(0,r_0^n)$, i.e., there exists $f_n(t) \in A(t)u_n(t)$ for almost everywhere $t \in [0,T]$ such that

$$(u_n', v) + (f_n, v)dt + \frac{1}{n}(Ju_n, v) = 0 \quad \text{for all } v \in W^{1,p}(0,T;E). \tag{6.7.4}$$

Put $v = u_n$ in (6.7.4). Then, by the condition (3), we know that there exists $N > 0$ such that

$$\|u_n\|_p \leq N, \quad \frac{1}{\sqrt{n}}\|u_n'\|_{*,q} \leq N \quad \text{for all } n \geq 1.$$

By the condition (2), u_n' is bounded in $L^q(0,T;E^*)$.

Now, we may assume that $u_n \rightharpoonup u_0$ in $W^{1,p}(0,T;E)$. Again, by the condition (2), we know that (f_n) is bounded in $L^q(0,T;E^*)$, and so we may assume that $f_n \rightharpoonup f_0$ in $L^q(0,T;E^*)$. Let $v = u_n - u_0$. It follows from (6.7.4) that

$$\lim_{n\to\infty}\int_0^T (f_n(s), u_n(s) - u_0(s))ds = 0.$$

Since $A(t)$ is a mapping of class (S_+), it is easy to show that $(f_n(s), u_n(s) - u_0(s)) \to 0$ in measure and hence there exists a subsequence $((f_{n_k}, u_{n_k} - u_0))$ such that $(f_{n_k}(s), u_{n_k}(s) - u_0(s)) \to 0$ for almost everywhere $s \in [0,T]$. Thus $u_{n_k}(s) \to u_0(s)$ for almost everywhere $s \in [0,T]$ and f_{n_k} converges weakly to $f_0(s) \in A(s)u(s)$. By letting $n_k \to \infty$ in (6.7.4), we get

$$u_0'(s) \in A(s)u(s) \quad \text{for almost everywhere } s \in (0,T).$$

It is obvious that $u_0(0) = u_0(T)$. This completes the proof.

Example 6.7.3. Let $\Omega \subset R^N$ be a bounded domain with a smooth boundary $\partial\Omega$. Let $a_i(x,u,\xi) : \Omega \times R \times R^N \to R$ be continuous with respect to u, ξ and measurable with respect to x for $1 \leq i \leq N$. Suppose that

(1) $|a_i(x,u,\xi)| \leq L(|u|^{m_1} + \|\xi\|^{m-1}) + M(x)$ for $1 \leq i \leq n$, where $L > 0$, $m_1 < \frac{N}{N-m}$, $2 \leq m < N$ are constants and $M(x) \in L^{\frac{m}{m-1}}(\Omega)$;

(2) ρ is continuous on R and

$$0 \leq \rho(u) \leq \mu\Big(\| \int_0^u \rho(s)ds\| + 1\Big)^r \quad \text{for all } u \in R,$$

where $\mu > 0$ and $0 \leq r < \frac{N}{N-2}$ are constants;

(3) $\Sigma_{i=1}^n [a_i(x,u,\xi) - a_i(x,u,\eta)](\xi_i - \eta_i) \geq k\|\xi - \eta\|^m$;

(4) $f(x,z) : \Omega \times R \to R$ satisfy the following conditions:

(4a) There exists $\Omega_0 \subset \Omega$ with $mes(\Omega_0) = 0$ such that

$$D_f = \bigcup_{x \notin \Omega_0} \{z \in R : f(x,.) \text{ is discontinuous at z}\}$$

has measure zero;

(4b) $x \to f(x,z)$ is measurable for all $z \notin D_f$;

(4c) There exist $k > 0$ and $l > 0$ such that

$$|f(x,z)| \leq k|z|^{m-1} + l \quad \text{for all } z \in D_f^c,\ a.e. x \in \Omega.$$

We consider the boundary value problem with discontinuous nonlinearity:

$$\begin{cases} \Sigma_{i=1}^N \frac{\partial}{\partial x_i}[\rho^2(u)\frac{\partial u}{\partial x_i} + a_i(x,u,\frac{\partial u}{\partial x})] + f(x,u) = 0, & a.e. x \in \Omega. \\ u(x) = 0, & x \in \partial\Omega. \end{cases} \qquad (E\ 6.7.2)$$

Choose $\Omega_1 \subset \Omega$ such that $mes\Omega_1 = 0$ and (4c) holds for all $x \in \Omega_1$. By (4a), there exists a countable subset $D \subset D_f^c$ satisfying $\overline{D} = R$. For all $(x,z) \in \Omega \times R$, we define

$$F(x,z) = \begin{cases} \bigcap_{k \in \mathcal{N}} co(\overline{f(x, [z-k^{-1}, z+k^{-1}] \cap D)}), & x \notin \Omega_1, \\ R, & \text{otherwise.} \end{cases}$$

It is known that $F(x,z)$ is nonempty closed convex and compact for all $x \in \Omega_1^c$ (see [58]). Moreover, $F(x,z)$ is measurable in x and $z \to F(x,z)$ is upper semicontinuous for almost all $x \in \Omega$.

Next, we define a mapping $\Psi : W_0^{1,m}(\Omega) \to 2^{L^q(\Omega)}$ by

$$\Psi(u) = \{v \in L^q(\Omega) : v(x) \in F(x,u(x)),\ a.e. x \in \Omega\}$$

for all $u \in W_0^{1,m}(\Omega)$, where $\frac{1}{m} + \frac{1}{q} = 1$. By Chen, Cho, and Yang [58], we know that $\Psi(u)$ is well defined and closed convex and the following properties hold:

(i) $u_n \to u \in W_0^{1,p}(\Omega)$ and $v_n \in \Psi^t(u_n)$, $v_n \rightharpoonup v$ in $L^q(\Omega)$ imply that $v \in \Psi(t,u)$;

(ii) Ψ maps bounded subsets of $W_0^{1,p}(\Omega)$ to weakly relatively compact subsets of $L^q(\Omega)$.

Finally, we define a mapping $A : W_0^{1,m}(\Omega) \to W^{-1,m}$ by

$$(Au, v) = \int_\Omega \Big\{\Sigma_{i=1}^n \Big[\rho^2(u)\frac{\partial u}{\partial x_i} + a_i\Big(x, u, \frac{\partial u}{\partial x}\Big)\Big]\frac{\partial v(x)}{\partial x_i}\Big\}dx$$

for all $u \in D(A)$ and $v \in W_0^{1,m}(\Omega)$, where

$$D(A) = \Big\{u \in W_0^{1,m}(\Omega) : \rho^2(u)\frac{\partial u}{\partial x} \in L^{\frac{m}{m-1}}\Big\}.$$

Since $F(x,z) = \{f(x,z)\}$ for all $x \in \Omega_1^c$, $z \in D_f^c$, the equation $(E\ 6.7.2)$ is equivalent to the following equation:

$$0 \in Au + \Psi u, u \in W_0^{1,m}(\Omega).$$

Theorem 6.7.4. $A + \Psi$ is a mapping of class $(S_+)_{0,L}$, where $L = C_0^\infty(\Omega)$.

Proof. Let $(F_j)_{j=1}^\infty$ be a sequence of finite dimensional subspace of $W_0^{1,m}(\Omega)$ such that $L \subset \overline{\cup_{j=1}^\infty F_j}$. Let $(u_j) \subset L$ satisfy

$$\limsup_{j\to\infty}(g_j, u_j) \le 0, \lim_{j\to\infty}(g_j, v) = 0 \quad \text{for all } v \in \cup_{j=1}^\infty F_j,$$

where $g_j \in Au_j + \Psi u_j$. Let $g_j = Au_j + f_j$, where $f_j \in \Psi u_j$. We have

$$\begin{aligned} &k\int_\Omega |\frac{\partial(u_j - v)}{\partial x}|^m dx \\ &\le \sum_{i=1}^N \int_\Omega [a_i(x, u_j, \frac{\partial u_j}{\partial x}) - a_i(x, u_j, \frac{\partial v}{\partial x})](\frac{\partial u_j}{\partial x} - \frac{\partial v}{\partial x})dx \\ &= (g_j, u_j - v) - (f_j, u_j - v) \\ &\quad -\sum_{i=1}^N \int_\Omega [\rho^2(u_j)\frac{\partial u_j}{\partial x} + a_i(x, u_j, \frac{\partial v}{\partial x})](\frac{\partial u_j}{\partial x} - \frac{\partial v}{\partial x})dx. \end{aligned} \tag{6.7.5}$$

Put $v(x) \equiv 0$. Then, by (1) and (4c), there exists a constant $M > 0$ such that

$$\limsup_{j\to\infty} \int_\Omega \rho^2(u_j)|\frac{\partial u_j}{\partial x}|^2 dx \le M. \tag{6.7.6}$$

By (6.7.6), we may also assume that

$$\lim_{j\to\infty} \int_\Omega \rho^2(u_j)|\frac{\partial u_j}{\partial x}|^2 dx = M_1 \tag{6.7.7}$$

by taking a subsequence, where M_1 is a constant. Set

$$\tilde{\rho}(u) = \int_0^u \rho(s)ds, \quad \tilde{u_j}(x) = \tilde{\rho}(u_j(x)).$$

Then $\tilde{u_j} \in W_0^{1,2}(\Omega)$. We may assume that $(\tilde{u_j})$ converges strongly in $L^p(\Omega)$, $p < \frac{2N}{N-2}$, to some $\tilde{u_0}$. Otherwise, take a subsequence. From this, we obtain that $\tilde{u_j}$ converges in measure to $\tilde{u_0}$ and $\widetilde{\rho(u_0)}$. Consequently, we have

$$\tilde{u_0}(x) = \widetilde{\rho(u_0)}(x). \tag{6.7.8}$$

By the assumption (2) and the boundedness of $(\tilde{u_j})$ in $L^{\frac{2N}{2N-2}}(\Omega)$, we get $(\rho(u_j))$ is bounded in $L^{\frac{2N}{2N-2}}(\Omega)$. Since $\rho(u_j)$ converges in measure, we obtain $\rho(u_j) \to \rho(u_0)$ in $L^2(\Omega)$.

Now, we can pass to the limit in (6.7.5) for all $v \in \cup_{j=1}^{\infty} F_j$ to obtain

$$\begin{aligned} & k\int_\Omega |\frac{\partial(u_j - v)}{\partial x}|^m dx \\ & \leq -M_1 - (f_0, u_0 - v) + \Sigma_{i=1}^N \rho^2(u_0)\frac{\partial u_0}{\partial x_i}\frac{\partial v}{\partial x_i}dx \\ & - \sum_{i=1}^N \int_\Omega a_i(x, u_0, \frac{\partial v}{\partial x})\frac{\partial(u_0 - v)}{\partial x_i}dx \leq -(f_0, u_0 - v) \\ & - \sum_{i=1}^N \int_\Omega [\rho^2(u_0)\frac{\partial u_0}{\partial x_i} + a_i(x, u_0, \frac{\partial v}{\partial x})]\frac{\partial(u_0 - v)}{\partial x_i}dx, \end{aligned} \tag{6.7.9}$$

where $f_0 \in \Psi u_0$. By taking limits in (6.7.9), we know that (6.7.9) is also true for all $v \in W_0^{1,q}(\Omega)$ for $q = 2[1 - \frac{N-2}{N}r]^{-1}$, where r is the same as in the assumption (2).

Now, consider the following functional on $W_0^{1,q}(\Omega)$ defined by

$$h(\phi) = \Sigma_{i=1}^N \int_\Omega \rho^2(u_0)\frac{\partial u_0}{\partial x_i}\frac{\partial \phi}{\partial x_i}dx. \tag{6.7.10}$$

It follows from (6.7.9) that h is bounded on $W_0^{1,q}(\Omega)$. Consequently, h can be extended to a linear functional on $W_0^{1,m}(\Omega)$ and we denote this extension by $\tilde{h}$. Notice that the Laplace operator $\Delta : W_0^{1,m'}(\Omega) \to (W_0^{1,m}(\Omega))^*$, $m' = \frac{m}{m-1}$, is a homeomorphism and there exists $u' \in W_0^{1,m'}(\Omega)$ such that

$$\tilde{h}(\phi) = \Sigma_{i=1}^N \int_\Omega \frac{\partial u'}{\partial x_i}\frac{\partial \phi}{\partial x_i}dx \quad \text{for all } \phi \in W_0^{1,m}(\Omega). \tag{6.7.11}$$

From (6.7.10) and (6.7.11), we get

$$\Sigma_{i=1}^N \int_\Omega \frac{\partial(u''(x) - u'(x))}{\partial x_i}\frac{\partial \phi}{\partial x_i}dx = 0 \quad \text{for all } \phi \in W_0^{1,m}(\Omega), \tag{6.7.12}$$

where $u''(x) = \int_0^{u_0(x)} \rho^2(s)ds$. Thus $u''(x) = u'(x)$ and, consequently, we have

$$\rho^2(u_0)\frac{u_0(x)}{\partial x_i} = \frac{u'(x)}{\partial x_i} \in L^{m'}(\Omega). \tag{6.7.13}$$

Therefore $u_0 \in D(A)$. Notice that (6.6.9) is also true for all $v \in W_0^{1,m}(\Omega)$. Put $v(x) = u_0(x) + tw(x)$ for all $t > 0$ and $w \in W_0^{1,m}(\Omega)$ in (6.7.9). Then, divide the resulting inequality by t and pass to the limit. We obtain $0 \in Au_0 + \Psi u_0$.

Finally, we prove that $u_j \to u_0 \in W_0^{1,m}(\Omega)$. We may assume that

$$a_i(x, u_j(x), \frac{u_j(x)}{\partial x}) \rightharpoonup b_i(x) \text{ in } L^{m'}(\Omega).$$

Take $u'_j \in \cup_{j=1}^{\infty} F_j$ such that $u'_j \to u_0$ in $W_0^{1,m}(\Omega)$. By using the assumption (3), we have

$$\begin{aligned}
&k\int_\Omega |\frac{\partial(u_j - u'_j)}{\partial x}|^m dx \\
&\leq \Sigma_{i=1}^N \int_\Omega [a_i(x, u_j, \frac{\partial u_j}{\partial x}) - a_i(x, u_j, \frac{\partial u'_j}{\partial x})](\frac{\partial u_j}{\partial x} - \frac{\partial u'_j}{\partial x})dx \\
&= (g_j, u_j) - \int_\Omega \rho^2(u_j)|\frac{\partial u_j}{\partial x}|^2 dx - (f_j, u_j) \\
&- \Sigma_{i=1}^N \int_\Omega a_i(x, u_j, \frac{\partial u_j}{\partial x})\frac{\partial u'_j}{\partial x} dx \\
&- \Sigma_{i=1}^N \int_\Omega a_i(x, u_j, \frac{\partial u'_j}{\partial x})\frac{\partial(u_j - u'_j)}{\partial x} dx.
\end{aligned} \tag{6.7.14}$$

By letting $j \to \infty$ in (6.7.14), we get

$$\begin{aligned}
&k\int_\Omega |\frac{\partial(u_j - u'_j)}{\partial x}|^m dx \\
&\leq -(f_0, u_0) - \int_\Omega \rho^2(u_0)|\frac{\partial u_0}{\partial x}|^2 dx - \Sigma_{i=1}^N \int_\Omega h_i(x)\frac{\partial u_0(x)}{\partial x_i} dx.
\end{aligned} \tag{6.7.15}$$

Now, using the fact $\lim_{j\to\infty}(Au_j + f_j, v) = 0$ for all $v \in \cup_{j=1}^{\infty} F_j$, we obtain

$$(f_0, v) + \Sigma_{i=1}^N \int_\Omega [\rho^2(u_0)\frac{\partial u_0(x)}{\partial x_i} + h_i(x)]\frac{\partial v(x)}{\partial x_i} dx = 0 \tag{6.7.16}$$

for all $v \in \cup_{j=1}^{\infty} F_j$. From (6.7.16) and (6.7.15), it follows that $u_j \to u_0 \in W_0^{1,m}(\Omega)$. This completes the proof.

From Theorem 6.7.4, we know that, if for some open bounded subset $U \subset W_0^{1,m}(\Omega)$ and $0 \notin Au + \Psi u$ for all $u \in \partial U \cap D(A)$, then $deg(A + \Psi, U \cap D(A), 0)$ is well defined.

Corollary 6.7.5. If $deg(A + \Psi, U \cap D(A), 0) \neq 0$, then the problem ($E$ 6.7 2) has a solution.

Remark. Some results on degree theory in this chapter can also be given in locally convex space.

6.8 Exercises

1. Let $A : D(A) \subseteq E \to 2^{E^*}$ be a set-valued mapping of class (S_+) and $P : E \to 2^{E^*}$ be a bounded pseudomonotone mapping. Show that $P + A$ is a mapping of class (S_+).

2. Let $A : D(A) \subseteq E \to 2^{E^*}$ be a multi-valued mapping of class (S_+), $T : D(A) \to 2^{E^*}$ be an upper semicontinuous operator with closed convex values and T maps each bounded subset of $D(A)$ into a relatively compact subset of E^*. Show that $T + A$ is a multi-valued mapping of class (S_+).

3. Let $A : D(A) \subseteq E \to 2^{E^*}$ be a multi-valued mapping of class (S_+) and $M : D(M) \subseteq E \to 2^{E^*}$ be a maximal monotone operator. Show that $M_\lambda + A$ is a multi-valued mapping of class (S_+).

4. Let $M : D(M) \subseteq E \to 2^{E^*}$ be a maximal monotone operator and $P : E \to 2^{E^*}$ be a bounded pseudomonotone mapping. Suppose that

$$\lim_{\|x\| \to \infty} \frac{(g, x - x_0)}{\|x\|} = +\infty \quad \text{for some } x_0 \in D(M) \text{ and all } g \in Px.$$

   Show that $(M + P)(D(M)) = E^*$.

5. Let $\phi : D(\phi) \subseteq E \to \overline{R} = R \cup \{+\infty\}$ be a proper lower semicontinuous convex function and

$$\partial\phi(x) = \{f \in E^* : \phi(y) \geq \phi(x) + f(y - x) \text{ for all } y \in D(\phi)\}.$$

   Show that $\partial\phi$ is maximal monotone.

6. Let $\phi : D(\phi) \subseteq E \to \overline{R} = R \cup \{+\infty\}$ be a proper lower semicontinuous convex function and

$$\phi_\lambda(x) = \inf_{y \in D(\phi)} \left(\frac{1}{2\lambda}\|x - y\|^2 + \phi(y)\right).$$

   Show that $(\partial\phi)_\lambda = \partial\phi_\lambda$ and $\phi_\lambda(x) \to \phi(x)$.

7. Let $\phi : D(\phi) \subseteq E \to \overline{R} = R \cup \{+\infty\}$ be a proper lower semicontinuous convex function. Show that $\partial\phi$ is surjective and $(\partial\phi)^{-1}$ is bounded if and only if $\lim_{\|x\|\to\infty} \frac{\phi(x)}{\|x\|} = +\infty$.

8. Let $M : D(M) \subseteq E \to 2^{E^*}$ be a linear monotone operator. Show that M is single valued if $\overline{D(M))} = E$, and M is maximal monotone if and only if $Graph(M)$ is closed and M^* is monotone.

9. Let $M : D(M) \subseteq E \to 2^{E^*}$ be a maximal monotone mapping. Show that M is surjective if and only if M^{-1} is locally bounded.

10. Construct the degree theory for a bounded pseudomonotone mapping.

11. Construct the degree theory for a bounded generalized pseudomonotone mapping.

12. Construct the degree theory for the sum of a bounded pseudomontone mapping and a maximal monotone mapping.

13. Construct the degree theory for the sum of a bounded generalized pseudo-

    monotone mapping and a maximal monotone mapping.

14. Construct the degree theory for the sum of a maximal monotone mapping and an upper semicontinuous compact mapping with closed convex values.

15. Let H be a real Hilbert space and $L : D(L) \subseteq H \to H$ be a linear, one to one maximal monotone mapping. Compute $deg(L, D(L) \cap B(0, r), 0)$ for all $r > 0$.

16. Let H be a real Hilbert space, $L : D(L) = H \to H$ be a linear one-to-one maximal monotone mapping and $\phi : D(\phi) \subseteq H \to \overline{R}$ be a proper lower semicontinuous convex function such that $\phi(0) = 0 \leq \phi(x)$ for all $x \in D(\phi)$. Compute $deg(L + \partial\phi, B(0, r), 0)$ for all $r > 0$.

Chapter 7

FIXED POINT INDEX THEORY

The study of non-negative solutions to the nonlinear equations, especially the ordinary or partial differential equations and integral equations, is a very important problem in nonlinear functional analysis. The non-negativity condition can be described by a closed convex subset P in a Banach space which satisfies $\lambda P \subset P$ for all $\lambda \geq 0$ and $P \cap -P = \{0\}$. We are interested in solving the equation $Tx = y$ in P. The fixed point index theory has proved to be a useful tool in studying such an equation.

It is our goal to introduce this theory in this chapter. Chapter 7 has six main sections.

Some introductory material on cones (normal, minihedral, etc.) is presented in Section 7.1.

In Section 7.2, we present a fixed point index for countably condensing mappings based on Dugundji's extension theorem and the Leray Schauder degree. Various properties of this index are given in Theorem 7.2.2.

Section 7.3 presents a variety of fixed point theorems for mappings defined in cones of Banach spaces (see, in particular, theorems 7.3.1, 7.3.3, 7.3.7, and 7.3.13).

The results in Section 7.3 are used in Section 7.4 to prove various fixed point results for perturbations of condensing maps (see Theorem 7.4.1 and 7.4.3).

In Section 7.5, we present a fixed point index for continuous generalized inward mappings.

Finally, some applications to integral and differential equations are given in Section 7.6.

7.1 Cone in Normed Spaces

We begin this section by first introducing the concept of partial order.

Definition 7.1.1. Let X be a nonempty set. There is an equivalent relation $\leq$ such that

(1) If $x \leq y, y \leq z$ for some $x, y, z \in X$, then $x \leq z$;

(2) $x \leq x$ for all $x \in X$;

(3) If $x \leq y$ and $y \leq x$ for some $x, y \in X$, then $x = y$.

Then $\leq$ is said to be a partial order on X and $(X, \leq)$ is said to be a partially ordered set.

Let $(X, \leq)$ be a partially ordered set and $M \subset X$. For any $x, y \in M$, if $x \leq y$ or $y \leq x$ holds, then we say M is a well-ordered set. If $x^* \in X$ satisfies $y \leq x^*$ (respectively, $x^* \leq y$) for all $y \in D$, then x^* is said to be an upper bound (respectively, lower bound) of M. If $x_0 \in X$ satisfies $x_0 \leq y$ for some $y \in X$ and $y = x_0$, then x_0 is said to be a maximal element of X. Similarly, if $x_0 \geq y$ for some $y \in X$ and $x_0 = y$, then x_0 is said to be a minimal element of X.

Theorem 7.1.2. (**Zorn Lemma**) If every well-ordered set in X has an upper bound (lower bound), then X has a maximal element (minimal element).

A cone is a very useful concept that can be used to generate a partial order in a linear space. Usually, this method is easy to manipulate and has been widely used in searching for positive solutions of nonlinear equations (see [8], [176]). Now, we recall this concept as follows:

Definition 7.1.3. Let E be a linear vector space and P be a nonempty convex subset of E. Then P is called a cone if

(1) $\lambda x \in P$ for all $x \in P$ and $\lambda > 0$;

(2) $P \cap (-P) = \{0\}$.

If E is a linear space and $P \subset E$ is a cone, we define an *order* $\leq$ on E as follows:

$$x \leq y \text{ if and only if } y - x \in P.$$

Note that $\leq$ is a well-defined partial order on X.

Example 7.1.4. Let $P \subset R^n$ be given by $P = \{(x_1, x_2, \cdots, x_n) \in R^n : x_i \geq 0,\ i = 1, 2, \cdots, n\}$. Then P is a cone.

Example 7.1.5. Let Ω be a nonempty Lebesgue measurable set in R^n and $0 < m(\Omega) < \infty$. Suppose that $L^p(\Omega) = \{f(\cdot) : \Omega \to R$ is measurable and $\int_\Omega |f(x)|^p dx < \infty\}$, where $0 < p < 1$. Put

$$P = \{f \in L^p(\Omega) : f(x) \geq 0 \text{ for almost everywhere } x \in \Omega\}.$$

Then P is a cone in $L^p(\Omega)$.

The following well-known order principle is due to Brezis and Browder [30]:

Theorem 7.1.6. Let $(X, \leq)$ be a partially ordered set and $S(x) = \{y \in X : x \leq y\}$. Suppose that $\psi : X \to R$ is a function satisfying the following conditions:

(1) For $x \leq y$ with $x \neq y$, $\psi(x) \leq \psi(y)$;

(2) For $x_1 \leq x_2 \leq \cdots \leq x_n \leq \cdots$, if $\{\psi(x_n)\}$ bounded, then there exists $y \in X$ such that $x_n \leq y$;

(3) $\psi(S(x))$ is bounded for all $x \in X$.

Then, for all $x \in X$, there exists $z \in S(x)$ such that z is maximal in X.

Proof. For each $x \in X$, we prove that $\sup S(x)$ exists and it belongs to $S(x)$. To see this, let $F \subset S(x)$ be a well-ordered set. Set $\alpha = \sup_{x \in F} \phi(x)$. Then there exists a sequence $(y_n) \subset F$ such that $\phi(y_1) \leq \phi(y_2) \leq \cdots \to \alpha$. By the assumption (1), we must have $y_1 \leq y_2 \leq y_3 \leq \cdots$. By assumption (2), there exists $y \in X$ such that $y_n \leq y$. Notice that $x \leq y_n$, so we have $x \leq y$. Thus $y \in S(x)$ and $\phi(y) \geq \alpha$, so $y \geq z$ for all $z \in F$. Therefore, y is an upper bound of F in $S(x)$. By the Zorn Lemma, $\sup S(x)$ exists in $S(x)$. Assume that $z \in X$ such that $\sup S(x) \leq z$. Then $z \in S(x)$, so we must have $z = \sup S(x)$. Thus $\sup S(x)$ is a maximal element. This completes the proof.

Definition 7.1.7. Let X be a normed space, $P \subset X$ be a cone and $\leq$ be the partial ordering defined by P. Then

(1) P is said to be normal if $\inf\{x + y\| : x, y \in P, \|x\| = \|y\| = 1\} > 0$;

(2) P is said to be quasinormal if there exist $y \in P$ with $y \neq 0$ and $\sigma(y) > 0$ such that $\|x + y\| \geq \sigma(y)\|x\|$ for any $x \in P$.

We set $\sigma(y) = \inf\{\frac{\|x+y\|}{\|x\|} : x \in P, x \neq 0\}$ for $y \in P$ with $y \neq 0$ and define $\sigma = \sup\{\sigma(y) : y \in P, y \neq 0\}$, which is called the quasinormality of P. Then we have $\frac{1}{2} \leq \sigma \leq 1$ (see [180], [246]).

Let $P \subset E$ be a cone and $D \subset E$ be a nonempty set. If $y, z \in E$ satisfies the following:

(i) $x \leq y$ for all $x \in D$;

(ii) $x \leq z$ for all $x \in D$ and so $y \leq z$;

then y is said to be an supremum of D and we denote it by $\sup D$.

Similarly, if (i) $x \geq y$ for all $x \in D$, (ii) $x \geq z$ for all $x \in D$ and so $y \geq z$, then y is said to an infimum of D and we denote it by $\inf D$.

Definition 7.1.8. Let X be a normed space, $P \subset X$ be a cone and $\leq$ be the partial ordering defined by P. Then

(1) P is said to be reproducing if $P - P = X$ and total if $\overline{P - P} = X$;

(2) P is said to be regular if every increasing sequence which is bounded from above is already convergent, and is fully regular if every bounded increasing sequence is convergent;

(3) P is said to be lastering if there exist a cone P_1 and a constant $\alpha > 0$ such that $B(x, \alpha\|x\|) \subset P_1$ for all $x \in P \setminus \{0\}$;

(4) P is said to be minihedral if $\sup\{x, y\}$ exists for all $x, y \in X$ and strongly minihedral if every set which is bounded from above has a supremum.

Definition 7.1.9. Let X be a normed space, $P \subset X$ be a cone and $\leq$ be the partial ordering defined by P. Then

(1) The norm $\|\cdot\|$ is said to be monotonic if $0 \leq x \leq y$ implies that $\|x\| \leq \|y\|$;

(2) The norm $\|\cdot\|$ is said to be semimonotonic if $0 \leq x \leq y$ implies that $\|x\| \leq \alpha\|y\|$ for some $\alpha > 0$.

Proposition 7.1.10. Let X be a Banach space and $P \subset X$ be a cone. Then we have

(1) P is quasinormal;

(2) P is normal if and only if $\|\cdot\|$ is semimonotonic;

(3) If $P^0 \neq \emptyset$, then P is reproducing.

Proof. (1) Suppose that P is not quasinormal. Then, for any $y \in P$ with $\|y\| = 1$, there exist $x_n \in P$ such that

$$\|x_n + y\| < n^{-1}\|x_n\| \quad \text{for } n = 1, 2, \cdots.$$

Therefore, if $\|x_n\|$ is bounded, then $x_n + y \to 0$, i.e., $x_n \to -y \in P$, which is a contradiction. If $\|x_n\|$ is unbounded, then $\|x_n\|^{-1}x_n \to 0$, which is impossible. Therefore, P is quasinormal.

(2) If $\|\cdot\|$ is semimonotonic, then, for any $x, y \in P$ with $\|x\| = \|y\| = 1$, we have $\|x + y\| \geq \alpha^{-1}\|y\| = \alpha^{-1}$, and so P is normal. On the other hand, if P is normal and $\|\cdot\|$ is not semimonotonic, then there exist x_n and y_n such that $0 < x_n \leq y_n$ and $\|x_n\| > n\|y_n\|$ for $n = 1, 2, \cdots$. Put

$$v_n = \|x_n\|^{-1}x_n, \ w_n = \|y_n\|^{-1}y_n, \ z_n = \|n^{-1}w_n - v_n\|^{-1}(n^{-1}w_n - v_n).$$

Then we have $z_n \in P$ and $\|v_n + z_n\| \to 0$ as $n \to \infty$, i.e., P is not normal, which is a contradiction.

(3) Take $x_0 \in P^0$ and $r > 0$ such that $B(x_0, 2r) \subset P$. For any $x \in X$ with $x \neq 0$, we have $x_0 + r\|x\|^{-1}x \in P$. Moreover,

$$x = \|x\|r^{-1}(x_0 + r\|x\|^{-1}x) - \|x\|r^{-1}x_0 \in P - P,$$

thus P is reproducing. This completes the proof.

By Proposition 7.1.10, if P is normal, then we set

$$\sigma(P) = \min\{N: \text{ if } 0 \leq x \leq y, \text{ then } \|x\| \leq N\|y\|\},$$

which is called the normality of P.

The following result gives a relation between various cones:

Proposition 7.1.11. Let X be a Banach space and $P \subset X$ be a cone. Then we have the implications: P allows plastering $\Rightarrow$ P is fully regular $\Rightarrow$ P is regular $\Rightarrow$ P is normal.

Proof. (1) Assume that P_1 is a plastering cone of P. Take $x_0 \notin P_1$; then there exists $r_0 > 0$ such that $\overline{B(x_0, r_0)} \cap P_1 = \emptyset$. By Mazur's separation theorem for convex sets, there exist $f \in X^*$ and $\beta \in R$ such that $f(y) \geq \beta$ for all $y \in P_1$ and $f(y) < \beta$ for all $y \in \overline{B(x_0, r_0)}$. This and $\lambda x \in P_1$ for all $\lambda > 0$ and $x \in P_1$ imply that $f(y) > 0$ for all $y \in P_1$. But $B(x, \alpha\|x\|) \subset P_1$ for all $x \in P$, so we have $f(x) \geq \alpha\|x\|\|f\|$.

Assume that $x_1 \leq x_2 \leq \cdots \leq x_n \cdots$ and $(x_n)_{n=1}^{\infty}$ is bounded. Then we have

$$f(x_{n+m} - x_n) \geq \alpha\|x_{n+m} - x_n\|\|f\|.$$

Thus $(x_n)_{n=1}^{\infty}$ is a Cauchy sequence and, consequently, $(x_n)_{n=1}^{\infty}$ is convergent.

(2) To show that P is fully regular implies that P is regular, we first show that P is fully regular implies that P is normal. Assuming this is false, there exist $x_n, y_n \in P \cap \partial B(0,1)$ such that $\|x_n + y_n\| \leq \frac{1}{2^n}$ for $n = 1, 2, \cdots$. Put

$$z_{2n} = \Sigma_{i=1}^{2n}(x_i + y_i), \quad z_{2n+1} = z_{2n} + x_{2n+1},$$

then (z_n) is increasing and bounded, so (z_n) is convergent, which contradicts $\|z_{2n+1} - z_{2n}\| = 1$.

Now, suppose that $x_1 \leq x_2 \leq \cdots \leq x_n \leq y$. Then $y - x_n \leq y - x_1$ for $n > 1$, so we have $\|y - x_n\| \leq \alpha\|y - x_1\|$ for some $\alpha > 0$. Thus (x_n) is convergent.

(3) Assume that P is regular, but P is not normal. Then there exist $x_n, y_n \in P \cap \partial B(0,1)$ such that $\|x_n + y_n\| \leq \frac{1}{2^n}$ for $n = 1, 2, \cdots$. Then we have

$$z_n = \Sigma_{i=1}^{n} x_i \leq \Sigma_{i=1}^{\infty}(x_i + y_i) \in P$$

and (z_n) is increasing, so (z_n) is convergent, which contradicts $\|z_n - z_{n-1}\| = \|x_n\| = 1$. Thus P is regular $\Rightarrow$ P is normal. This completes the proof.

Proposition. 7.1.12. Let X be a normed space and $P \subset E$ be a regular cone. Then we have the following:

(1) If $D \subset E$ is a well-ordered set and has an upper bound, then $\sup D$ exists.

(2) If $D \subset E$ is a well-ordered set and has a lower bound, then $\inf D$ exists.

Proof. (2) follows from (1) by replacing D by $-D$.

Now, we prove (1). By Proposition 7.1.11, E is normal. Let x_0 be an upper bound of D. Define a function $\phi : D \to [0, +\infty)$ as follows:

$$\phi(x) = \sup\{\|z - y\| : x \leq y \leq z,\ y, z \in D\}.$$

Note that $z - y \leq x_0 - x$ and we have $\phi(x) \leq \sigma(P)\|x_0 - x\|$. Obviously, if $x \leq y$, then $\phi(x) \geq \phi(y)$. We claim that $\inf_{x \in D} \phi(x) = 0$. If not, $\inf_{x \in D} \phi(x) = \delta > 0$, there exists $x_1 \in D$ such that $\phi(x_1) > \delta$, so there are $x_1 \leq y_1 \leq z_1$ such that $\|z_1 - y_1\| > \delta$. Again, since $\phi(z_1) > \delta$, there are $z_1 \leq y_2 \leq z_2$ such that $\|z_2 - y_2\| > \delta$. Repeat the above process, we get a sequence (y_n) in D such that

$$y_1 \leq z_1 \leq y_2 \leq z_2 \leq \cdots \leq x_0$$

with $\|z_i - y_i\| > \delta$ for $i = 1, 2, \cdots$, so it does not converge, which contradicts the fact that P is regular. Thus there exist $(x_n)_{n=1}^{\infty} \subset D$ such that $\phi(x_n) \to 0$. Since D is well ordered, put $y_n = \max\{x_1, x_2, \cdots, x_n\}$. Then $y_1 \leq y_2 \leq \cdots \leq x_0$ and so $\phi(y_n) \to 0$. Since P is regular, there exists $y_0 \in E$ such that $y_n \to y_0$. It is easy to see that $y_0 = \sup\{y_i : i = 1, 2, \cdots\}$.

Next, we prove that $y_0 = \sup D$. To see this, we only need to prove that y_0 is an upper bound of D. For any $x \in D$, we have two cases: (1) $x \leq y_n$ for some n and (2) $y_n \leq x$.

In case (1), we have $x \leq y_0$.

In case (2), we have $y_0 \leq x$ and $0 \leq x - y_0 \leq x - y_n$ and thus

$$\|x - y_0\| \leq N\|y_n - x\| \leq N\phi(y_n) \to 0 \quad \text{as } n \to \infty.$$

Thus $x = y_0$. Combine (1) and (2), then we have $x \leq y_0$, so y_0 is an upper bound of D. Thus $y_0 = \sup D$. This completes the proof.

Proposition 7.1.13. Let $P \subset E$ be a cone. If P is regular and minihedral, then P is strongly minihedral.

Proof. Let $D \subset E$ and D has an upper bound x_0. Set

$$U = \{x : x \text{ is an upper bound of } D\},$$

then $x_0 \in U$. Consider any well-ordered set F of U. Obviously, any element in D is a lower bound for F. By Proposition 7.1.12, we know that $f = \inf F$ exists and $x \leq f$ for all $x \in D$, and thus $f \in U$. By the Zorn Lemma, U has a minimal element x^*.

Now, we prove that $x^* = \sup D$. To see this, we only need to prove $x^* \leq y$ for all $y \in U$. For any $y \in U$, put $z = \inf\{x^*, y\}$. We have $z \leq x^*, z \leq y$, but $x \leq x^*, x \leq y$. Thus we must have $x \leq z$ and so $z \in U$. Since x^* is minimal, we must have $z = x^*$ and hence $x^* \leq y$, as desired. This completes the proof.

Example 7.1.14. Let $E = l^{\infty}$ and $P = \{(x_i) \in l^{\infty} : x_i \geq 0,\ i = 1, 2, \cdots\}$. Then P is normal. For any set D with a upper bound, it is easy to see that $\sup D = z = (z_1, z_2, \cdots)$, where $z_i = \sup_{x \in D} x(i)$ and $x(i)$ is the i-th coordinate of x for $i = 1, 2 \cdots$. Thus, P is strongly minihedral. For any integer $n \geq 1$, let $x^n \in l^\infty$ be defined by

$$x^n(i) = \begin{cases} 1, & \text{if } i \leq n, \\ 0, & \text{if } i > n. \end{cases}$$

Then $x^1 \leq x^2 \leq \cdots \leq (1, 1, 1, \cdots, 1, \cdots)$. However, (x^n) does not converge in l^∞. Thus, P is not regular.

Proposition 7.1.15. Let E be a Banach space, $P \subset E$ be a cone and $B = \{x : \|x\| \leq 1\}$. The the following conclusions are equivalent:

(1) P is reproducing;

(2) There exists $r > 0$ such that, for any $x \in E$, there exist $y, z \in P$ with $\|y\| \leq r\|x\|$, $\|z\| \leq r\|x\|$ and $x = y - z$;

(3) There exists $\alpha > 0$ such that $\alpha B \subset B \cap P - B \cap P$;

(4) There exists $\beta > 0$ such that $\beta B \subset \overline{B \subset B \cap P - B \cap P}$.

Proof. It is obvious that (2) is equivalent to (3), (2) implies (1), and (3) implies (4), so we only need to prove (1) implies (2) and (4) implies (3).

(1) $\Rightarrow$ (2) Since P is reproducing, we have $E = \cup_{n=1}^{\infty} E_n$, where

$$E_n = \{x \in E : \text{ there exists } y \in P \text{ such that } x \leq y, \ \|y\| \leq n\|x\|\}$$

for $n = 1, 2, \cdots$. By Baire's Theorem, there exist an integer $n_0 > 0$, $x_0 \in E$, $r > s > 0$ such that

$$T = \{x \in E : s < \|x - x_0\| < r\} \subset E_{n_0}.$$

Let $y_0, z_0 \in P$ be such that $-x_0 = y_0 - z_0$. Take an integer $n_1 > 0$ such that $\|y_0\| \leq n_1\|x_0\|$. Set $T_0 = \{x \in E : s \leq \|x\| < r\}$. Take an integer n_2 such that

$$n_2 > n_0 + \frac{1}{s}(n_0 + n_1)\|x_0\|.$$

We prove that $T_0 \subset \overline{E_{n_3}}$. In fact, for any $x \in T_0$, we have $y + x_0 + x \in T$, so there exist $x_i \in E_{n_0}$, $i = 1, 2, \cdots$, such that $x_i \to y$. We also assume that $x_i \in T$. There exist $y_i \in P$ such that $x_i \leq y_i$ and $\|y_i\| \leq n_0\|x_i\|$ for $i = 1, 2, \cdots$. We also have $x_i - x_0 \leq y_i + y_0 \in P$ and

$$\begin{aligned} \|y_i - y_0\| &\leq n_0\|x_i\| + n_1\|x_0\| \leq (n_0 + n_1)\|x_0\| + n_0\|x_i - x_0\| \\ &\leq [(n_0 + n_1)\frac{\|x_0\|}{s} + n_0]\|x_i - x_0\| \leq n_2\|x_i - x_0\|. \end{aligned}$$

Therefore, $x_i - x_0 \in E_{n_2}$ for $i = 1, 2, \cdots$. Obviously, $x_i - x_0 \to y - x_0 = x$ as $i \to \infty$, so $x \in \overline{E_{n_2}}$.

Finally, we prove that $E = E_{3n_2}$. For any $x \in E \setminus \{0\}$, take $x_1 \in E_{n_2}$ such that

$$\|x - x_1\| < \frac{1}{2}\|x\|.$$

There exist $y_1 \in P$ such that $x_1 \leq y_1$ and $\|y_1\| \leq n_2\|x_1\|$. Similarly, there exist $x_2 \in E_{n_3}$ and $y_2 \in P$ such that

$$\|x - x_1 - x_2\| < \frac{1}{2^2}\|x\|, x_2 \leq y_2, \|y_2\| \leq n_2\|x_2\|.$$

Repeating this process, we get two sequences $(x_k)_{k=1}^{\infty} \subset E_{n_2}$, $(y_k)_{k=1}^{\infty} \in P$ such that

$$\|x - x_1 - x_2 - \cdots - x_k\| < \frac{1}{2^k}\|x\|, x_k \leq y_k, \|y_k\| \leq n_2\|x_k\|$$

for $k = 1, 2, \cdots$. Obviously, $x = \Sigma_{i=1}^{\infty} x_i$ and $\|x_k\| < \frac{3\|x\|}{2^k}$ for $k = 1, 2, \cdots$. Thus, $\Sigma_{i=1}^{\infty}\|y_i\| \leq 3n_2\|x\|$. Put $y = \Sigma_{i=1}^{\infty} y_i$; then $x \leq y$ and $\|y\| \leq 3n_2\|x\|$. Thus, $x \in E_{3n_2}$.

(4) $\Rightarrow$ (3) Set $C = B \cap P - B \cap P$. Let $\beta > 0$ be such that $\beta B \subset \overline{C}$. We prove that $\frac{\beta}{2}B \subset C$. For any $y \in \frac{\beta}{2}B$, there exists $2y_1 \in C$ such that $\|2y - 2y_1\| < \frac{\beta}{2}$. Again, there exists $2^2 y_2 \in C$ such that $\|2^2 y - 2^2 y_1 - 2^2 y_2\| < \frac{\beta}{2}$. Repeat this process, we get a sequence $(y_n)_{n=1}^{\infty} \subset E$ such that $2^n y_n \in C$ and

$$\|y - y_1 - y_2 - \cdots - y_n\| < \frac{\beta}{2^{n+1}} \quad \text{for } n = 1, 2, \cdots .$$

Thus, we have

$$y = \sigma_{i=1}^{\infty} y_i, \; y_n = x_n - z_n, \; x_n, z_n \in P, \; \|x_n\| \leq \frac{1}{2^n}, \; \|z_n\| \leq \frac{1}{2^n}$$

for $n = 1, 2, \cdots$. Put $x = \Sigma_{i=1}^{\infty} x_i$, $z = \Sigma_{i=1}^{\infty} z_i$. Then $x, z \in P$, $\|x\| \leq 1$, $\|z\| \leq 1$ and $x \in C$. Thus, $\frac{\beta}{2}B \subset C$. This completes the proof.

7.2 Fixed Point Index Theory

In this section, let E be a Banach space and $P \subset E$ be a cone.

Lemma 7.2.1. [291] Let $B \subset P$ be a bounded closed subset and $T : B \to P$ a countably condensing mapping. Set $C_1 = \overline{conv(TB)}$, $C_{n+1} = \overline{conv(T(C_n \cap B))}$ for $n \geq 1$ and $C = \cap_{n=1}^{\infty} C_n$. If $M \subset E$ and $M \setminus C_n$ is finite for $n = 1, 2, \cdots$, then M is relatively compact. In particular, C is compact.

Proof. Let $\mathcal{F}$ be the family of all subsets M such that $M \setminus C_n$ is finite for $n = 1, 2, \cdots$ and $\mathcal{F}_B$ be the family of all countable subsets $M \in \mathcal{F}$ with $M \subset B$.

Step 1 We prove that there exists $B^* \in \mathcal{F}_B$ such that $\alpha(K) \leq \alpha(B^*)$ for all $K \in \mathcal{F}_B$. In fact, since $\alpha(K) \leq \alpha(B)$ for all $K \in \mathcal{F}_B$, we have $s = \sup_{K \in \mathcal{F}_B} \alpha(K) < +\infty$, and let $K_n \in \mathcal{F}_B$ such that $\alpha(K_n) \to s$ as $n \to \infty$. Put $B^* = \cup_{n=1}^{\infty} K_n$, then B^* is countable and $\alpha(B^*) = s \geq \alpha(K)$ for all $K \in \mathcal{F}_B$.

Step 2 If $M \in \mathcal{F}$, and $x_n \in M$ for $n = 1, 2, \cdots$ and no x_n appears infinitely many times, then there exist $A \in \mathcal{F}_B$ and $y_n \in conv(TA)$ such that $\|x_n - y_n\| \to 0$ as $n \to \infty$. To see this, observe that $M \setminus C_1$ is finite and it contains at most finitely many x_n, and so we may assume that $x_n \in C_1$ for all $n \geq 1$. For any given integer n, let k_n be the largest integer such that $x_n \in C_{k_n}$. If no such integer exists, put $k_n = n$. For any integer k, $M \setminus C_k$ is finite, thus $\{n : x_n \notin C_k\}$ is finite. One can easily see that

$$I_k = \{n : k_n \leq k\} \subset I_{k+1} \cup \{1, 2, \cdots, k\}$$

by virtue of $C_1 \supset C_2 \supset \cdots$ and thus I_k is finite for all k. We have

$$x_n \in C_{k_n} = \overline{conv(T(C_{k_n-1} \cap B))} \quad \text{for any } n.$$

Hence there exists $y_n \in conv((T(C_{k_n-1} \cap B))$ such that $\|x_n - y_n\| < n^{-1}$. In particular, we can find some finite $A_n \subset C_{k_n-1} \cap B$ such that $y_n \in conv(TA_n)$.

Now, if we put $A = \cup_{n=1}^{\infty} A_n$, then A is the required subset. To see this, we have to check that $A \in \mathcal{F}_B$. Since $C_1 \supset C_2 \supset \cdots$, we have $A_n \subset C_i$ for $i \leq k_n - 1$, and thus we have

$$A \setminus C_n = \cup_{i=1}^{\infty}(A_i \setminus C_n) = \cup_{i;n>k_i-1}(A_i \setminus C_n) \subset \cup_{i \in I_n} A_i,$$

but the last set is finite, so we get the desired result.

Step 3 We prove that any $K \in \mathcal{F}_B$ is finite. For any $K \in \mathcal{F}_B$, if K is finite, we are done. So we assume that K is infinite. Replace K by $K \cup B^*$, we get $\alpha(K) = s$. The countability of K implies that $K = \{x_n : n \geq 1\}$.

By Step 2, there exist $A \in \mathcal{F}_B$ and $y_n \in conv(TA)$ such that $\|x_n - y_n\| \to 0$ as $n \to \infty$. Obviously, $K \cup A \in \mathcal{F}_B$ and thus $\alpha(A \cup K) = \alpha(K) = s$. Put

$$K_n = \{x_1, x_2, \cdots, x_n, y_{n+1}, y_{n+2}\}, \quad K_0 = \{y_1, y_2, \cdots\}.$$

We have $\alpha(K_0) = \alpha(K) = s$. On the other hand,

$$s = \alpha(K_0) \leq \alpha(conv(TA)) \leq \alpha(T(A \cup K)).$$

But T is countably compact, so we must have $s = 0$. Consequently, K is relatively compact.

Step 4 We prove that any $F \in \mathcal{F}$ is relatively compact. For any sequence $(x_n) \subset F$, if x_n appears infinitely many times for some n, we are done already. Otherwise, by Step 2, we may take y_n and A as in Step 2. By Step 3, A is relatively compact, so $conv(T(A)$ is relatively compact. Thus $\{y_i : i \geq 1\}$ is relatively compact and so F is relatively compact. This completes the proof.

Construction of Index Theory. Let $\Omega \subset E$ be an open bounded subset and $\overline{\Omega} \cap P \neq \emptyset$. Let $T : \overline{\Omega} \cap P \to P$ be a countably condensing and continuous mapping. Put

$$C_0 = \overline{conv(T(\overline{\Omega \cap P}))}, \quad C_n = \overline{conv(T(C_{n-1} \cap \overline{\Omega})} \quad \text{for } n \geq 1,$$

then, by Lemma 7.2.1, $C = \cap_{n=1}^{\infty} C_n$ is compact and $T : C \to C$ is a mapping. Now, assume also that $Tx \neq x$ for all $x \in \partial\Omega \cap P$. Then we have the following two cases:

Case (1) If $C = \emptyset$, we define $ind(T, \Omega \cap P) = 0$.

Case (2) If $C \neq \emptyset$ and $r : E \to C$ is a retraction, then $r^{-1}(\Omega \cap C)$ is open in E. It is easy to see that $Trx \neq x$ for all $x \in \partial(\Omega \cap r^{-1}(\Omega))$ and a mapping $Tr : \overline{\Omega \cap r^{-1}(\Omega)} \to C$, so the Leray Schauder degree $deg(I - Tr, \Omega \cap r^{-1}(\Omega), 0)$ is well defined. We define

$$ind(T, \Omega \cap P) = deg(I - Tr, \Omega \cap r^{-1}(\Omega), 0),$$

which is called the fixed point index of T.

A slight modification of the argument in Section 2 of Chapter 3 guarantees that $deg(I - Tr, \Omega \cap r^{-1}(\Omega), 0)$ does not depend on r, so $ind(T, \Omega \cap P)$ is well defined.

Theorem 7.2.2. The fixed point index $ind(T, \Omega \cap P)$ satisfying the following properties:

(1) $ind(x_0, \Omega \cap P) = 1$ if $x_0 \in \Omega \cap P$;

(2) If $ind(T, \Omega \cap P) \neq 0$, then $x = Tx$ has a solution in Ω;

(3) If $\Omega_i \subset \Omega$ for $i = 1, 2$, $\Omega_1 \cap \Omega_2 = \emptyset$ and $0 \notin (I - T)[(\overline{\Omega} \setminus (\Omega_1 \cup \Omega_2)) \cap P]$, then

$$ind(T, \Omega \cap P) = ind(T, \Omega_1 \cap P) + ind(T, \Omega_2 \cap P);$$

(4) If $H(t, x) : [0, 1] \times \overline{\Omega} \cap P \to P$ is a continuous mapping satisfying $\alpha(H([0, 1] \times B)) < \alpha(B)$ for any countable subset B of $\overline{\Omega} \cap P$ with $\alpha(B) \neq 0$ and $x \neq H(t, x)$ for all $(t, x) \in [0, 1] \times \partial\Omega$, then $ind(H(t, \cdot), \Omega \cap P)$ does not depend on $t \in [0, 1]$.

Proof. It is similar to the proof of Theorem 3.2.1.

Note that, in the case that $T : \overline{\Omega} \cap P \to P$ is continuous and compact and $x \neq Tx$ for all $x \in \partial\Omega \cap P$, one can define the fixed point index by

$$ind(T, \Omega \cap P) = deg(I - Tr, r^{-1}(\Omega \cap P) \cap \Omega, 0),$$

where $r : E \to P$ is a retraction. One can check easily this definition does not depend on r and we have the following properties:

Theorem 7.2.3. The fixed point index $ind(T, \Omega \cap P)$ satisfies the following properties:

(1) $ind(x_0, \Omega \cap P) = 1$ if $x_0 \in \Omega \cap P$;

(2) If $ind(T, \Omega \cap P) \neq 0$, then $x = Tx$ has a solution in Ω;

(3) If $\Omega_i \subset \Omega$ for $i = 1, 2$, $\Omega_1 \cap \Omega_2 = \emptyset$ and $0 \notin (I - T)[(\overline{\Omega} \setminus (\Omega_1 \cup \Omega_2)) \cap P]$, then

$$ind(T, \Omega \cap P) = ind(T, \Omega_1 \cap P) + ind(T, \Omega_2 \cap P);$$

(4) If $H(t, x) : [0, 1] \times \overline{\Omega} \cap P \to P$ is a continuous compact such that $x \neq H(t, x)$ for all $(t, x) \in [0, 1] \times \partial\Omega$, then $ind(H(t, \cdot), \Omega \cap P)$ does not depend on $t \in [0, 1]$.

7.3 Fixed Point Theorems in Cones

In this section, we derive some fixed point theorems by using fixed point index theory.

Theorem 7.3.1. Let E be a Banach space, P be a cone in E, Ω be an open bounded subset of E, $0 \in \Omega_1 \subset \Omega$ be an open subset of Ω, and $T : \overline{\Omega \setminus \Omega_1 \cap P} \to P$ be a completely continuous mapping. Suppose that one of the following conditions is satisfied:

(1) $Tx \geq x$ for all $x \in \partial\Omega \cap P$ and $Tx \leq x$ for all $x \in \partial\Omega_1 \cap P$;

(2) $Tx \geq x$ for all $x \in \partial\Omega_1 \cap P$ and $Tx \leq x$ for all $x \in \partial\Omega \cap P$.

Then T has a fixed point in $\overline{\Omega \setminus \Omega_1} \cap P$.

To prove Theorem 7.3.1, we first prove the following result:

Lemma 7.3.2. Suppose that $C \subset P$ is a compact set such that $0 \notin C$. Then $0 \notin Conv(C)$.

Proof. Since $0 \notin C$ and C is compact, there exists $\epsilon_0 > 0$ such that

$$d(0, C) \geq \epsilon_0.$$

We also have $C \subset \cup_{x \in C} B_x(\epsilon_0)$, so there exist finitely many x_i for $i = 1, 2, \cdots, n$ such that

$$C \subset \cup_{i=1}^n B_{x_i}(\epsilon_0).$$

Therefore, it follows that

$$Conv(C) = Conv(\cup_{i=1}^n B_{x_i}(\epsilon_0) \cap C).$$

If $0 \in Conv(C)$, then there exists $y_i \in B_{x_i}(\epsilon_0) \cap C$, $\alpha_i \in (0,1)$ for $i = 1, 2, \cdots, n$ and $\Sigma_{i=1}^n \alpha_i = 1$ such that

$$0 = \Sigma_{i=1}^n \alpha_i y_i,$$

which contradicts $y_i > 0$ for $i = 1, 2, \cdots, n$. This completes the proof.

Proof of Theorem 7.3.1. We may assume that (1) holds (the proof is similar when (2) holds). By our assumption, we have $0 \notin \overline{T(\partial\Omega \cap P)}$. In view of Lemma 7.3.2, it follows that

$$0 \notin Conv(\overline{T\partial\Omega \cap P})).$$

Hence, we can find a completely continuous mapping $T_1 : E \to conv(\overline{T\partial\Omega \cap P}))$ such that $T_1 x = Tx$ for all $x \in \partial\Omega \cap P$.

Now, for any $k > 1$, we have $tTx + (1-t)kT_1 x \neq x$ for all $(t, x) \in [0,1] \times \partial\Omega \cap P$, and so by Theorem 7.2.3

$$ind(kT_1, \Omega \cap P) = ind(T, \Omega \cap P) \quad \text{for all } k > 1.$$

Therefore, $ind(T, \Omega \cap P)$ must equal to 0 (otherwise, Ω is unbounded, which is a contradiction). It is easy to see that $ind(T, \Omega_1 \cap P) = 1$, and so we have

$$ind(T, \Omega \setminus \overline{\Omega_1} \cap P) = 1.$$

Therefore, T has a fixed point in $\overline{\Omega \setminus \Omega_1} \cap P$. This completes the proof.

By the same proof as in Theorem 7.3.1, we get the following more general result:

Theorem 7.3.3. Let P be a cone in a Banach space E, Ω be an open bounded subset of E, $\Omega_1 \subset \Omega$ be an open subset of Ω with $0 \in \Omega_1$, and $T : \overline{\Omega \setminus \Omega_1} \cap P \to P$ be a completely continuous mapping. Suppose that one of the following conditions is satisfied:

(1) $Tx \neq \lambda x$ for all $\lambda > 1$ and $x \in \partial\Omega \cap P$;

(2) $Tx \neq \alpha x$ for all $x \in \partial\Omega_1 \cap P$ and $\alpha < 1$;

(3) $\inf\{\|Tx\| : x \in \partial\Omega_1 \cap P\} > 0$.

Then T has a fixed point in $\overline{\Omega \setminus \Omega_1} \cap P$.

From Theorem 7.3.3, we have the following:

Corollary 7.3.4. Let P be a cone in a Banach space E, Ω be an open bounded subset of E, $\Omega_1 \subset \Omega$ be an open subset of Ω with $0 \in \Omega_1$ and $T : \overline{\Omega \setminus \Omega_1} \cap P \to P$ be a completely continuous mapping. Suppose that one of the following conditions is satisfied:

(1) $\|Tx\| \leq \|x\|$ for all $x \in \partial\Omega \cap P$;

(2) $\|Tx\| \geq \|x\|$ for all $x \in \partial\Omega_1 \cap P$.

Then T has a fixed point in $\overline{\Omega \setminus \Omega_1} \cap P$.

Another variant form of Theorem 7.3.3 is as follows:

Theorem 7.3.5. Let P be a cone in a Banach space E, Ω be an open bounded subset of E, $\Omega_1 \subset \Omega$ be an open subset of Ω with $0 \in \Omega_1$ and $T : \overline{\Omega \setminus \Omega_1} \cap P \to P$ be a completely continuous mapping. Suppose that one of the following conditions is satisfied:

(1) $Tx \not\geq x$ for all $x \in \partial\Omega \cap P$;

(2) $Tx \not\leq x$ for all $x \in \partial\Omega_1 \cap P$;

(3) $\inf\{\|Tx\| : x \in \partial\Omega_1 \cap P\} > 0$.

Then T has a fixed point in $\overline{\Omega \setminus \Omega_1} \cap P$.

Theorem 7.3.3 fails if we assume that T is a k-set contraction with $k < 1$. This can be seen with the following example:

Example 7.3.6. Let a mapping $T : l^2 \to l^2$ be defined by

$$T(x_1, x_2, \cdots) = \frac{1}{3}(0, x_1, x_2, \cdots) \text{ for all } (x_1, x_2, \cdots) \in l^2.$$

Now, T is obviously $\frac{1}{3}$-set contraction. Take $P = \{(x_i) : x_i \geq 0 \text{ for } i = 1, 2, \cdots\}$ and $\Omega = B(0, 2)$, $\Omega_1 = B(0, 1)$. Then T has no fixed point in $\overline{\Omega \setminus \Omega_1} \cap P$.

In the following, we provide sufficient conditions such that the conclusion of Theorem 7.3.3 still holds in the case of k-set contractions.

Theorem 7.3.7. Let X be a real Banach space,Ω be an open bounded subset of X with $0 \in \Omega$ and P a quasinormal cone in X. If $T : \overline{\Omega} \cap P \to P$ is a continuous countably condensing mapping and the following condition is satisfied:

$$\inf\left\{\frac{\|Tx\|}{\|x\|} : x \in \partial\Omega \cap P\right\} > \frac{1}{\sigma},$$

where σ is the quasinormality constant of P, then $ind(T, \Omega \cap P) = 0$.

Proof. For any $\epsilon > 0$, by the definition of the quasinormality constant of P, there exists $y_\epsilon \in P$ with $y_\epsilon \neq 0$ such that

$$\|x + \lambda y_\epsilon\| \geq (\sigma - \epsilon)\|x\| \quad \text{for all } x \in P.$$

By assumption, we may fix $\epsilon > 0$ such that

$$\inf\left\{\frac{\|Tx\|}{\|x\|} : x \in \partial\Omega \cap P\right\} > \frac{1}{\sigma - \epsilon}. \tag{7.3.1}$$

For any $m > 0$, we claim that $x - Tx \neq tmy_\epsilon$ for all $x \in \partial\Omega \cap P$ and $t \in [0, 1]$. If not, then there exist $x_0 \in \partial\Omega \cap P$, $t_0 \in [0, 1]$ such that $x_0 = Tx_0 + t_0 my_\epsilon$, and so we have

$$\|x_0\| = \|Tx_0 + t_0 my_\epsilon\| \geq (\sigma - \epsilon)\|Tx_0\|.$$

Hence we get

$$\frac{\|Tx_0\|}{\|x_0\|} \leq \frac{1}{\sigma - \epsilon},$$

which contradicts (7.3.1). By the homotopy property (4) of Theorem 7.2.3, we get

$$ind(T, \Omega \cap P) = ind(T + m_0 y_\epsilon, \Omega \cap P)$$

and thus, we must have

$$ind(T + my_\epsilon, \Omega \cap P) = 0$$

(since Ω is bounded and $T(E)$ is bounded for any countably bounded subset E of $\overline{\Omega} \cap P$). Consequently, $ind(T, \Omega \cap P) = 0$. This completes the proof.

Corollary 7.3.8. Let X be a real Banach space, Ω be an open bounded subset of X with $0 \in \Omega$ and P be a quasinormal cone in X. If $T : \overline{\Omega} \cap P \to P$ is a continuous countably k-set contraction and the following conditions are satisfied:

(1) $Tx \neq \alpha x$ for all $x \in \partial\Omega \cap P$ and $\alpha \in [0, 1]$;

(2) $\inf\left\{\frac{\|\mu Tx\|}{\|x\|} : x \in \partial\Omega \cap P\right\} > \frac{1}{\sigma}$, where $1 \leq \mu < \frac{1}{k}$ and σ is the quasinormality constant of P;

then $ind(T, \Omega \cap P) = 0$.

Proof. If we put $T_1 x = \mu Tx$ for all $x \in \overline{\Omega} \cap P$, then T_1 is countably condensing and T_1 satisfies all the conditions of Theorem 7.3.7. Thus, we have

$$ind(T_1, \Omega \cap P) = 0.$$

From the assumption (1) and the homotopy property (4) of Theorem 7.2.3, we know that

$$ind(T, \Omega \cap P) = ind(\mu T, \Omega \cap P).$$

Therefore, we have $ind(T, \Omega \cap P) = 0$. This completes the proof.

Corollary 7.3.9. Let X be a real Banach space, Ω be an open bounded subset of X with $0 \in \Omega$ and P be a quasinormal cone in X. If $T : \overline{\Omega} \cap P \to P$ is a continuous countably k-set contraction with $k < \sigma$, where σ is the quasinormality constant of P and the following conditions are satisfied:

(1) $Tx \neq x$ for all $x \in \partial\Omega \cap P$;

(2) $\|Tx\| \geq \|x\|$ for all $x \in \partial\Omega \cap P$;

then $ind(T, \Omega \cap P) = 0$.

Proof. We take $\epsilon_0 > 0$ such that $k + \epsilon_0 < \min\{1, \sigma\}$ and put $T_1 = \frac{1}{k+\epsilon_0}T$. Then T_1 is a countably $\frac{k}{k+\epsilon_0}$-contraction, and we know, by the assumption (2), that

$$\inf\left\{\frac{\|T_1x\|}{\|x\|} : x \in \partial\Omega \cap P\right\} \geq \frac{1}{k + \epsilon_0} > \frac{1}{\sigma} \text{ for all } x \in \partial\Omega \cap P.$$

We also have $T_1x \neq \alpha x$ for all $x \in \partial\Omega \cap P$ and $\alpha \in [0, 1]$. Therefore, by Theorem 7.3.7, we have $ind(T_1, \Omega \cap P) = 0$. This completes the proof.

Theorem 7.3.10. Let X be a real Banach space, Ω be an open bounded subset of X with $0 \in \Omega$ and P be a cone in X. Let $T : \overline{\Omega} \cap P \to P$ be a continuous countably condensing mapping. Suppose that there exists $y_0 \in P \setminus \{0\}$ such that $x - Tx \neq \lambda y_0$ for all $x \in \partial\Omega \cap P$ and $\lambda \geq 0$. Then $ind(T, \Omega \cap P) = 0$.

Proof. For any $m > 0$, by the homotopy property (4) of Theorem 7.2.3, we have

$$ind(T, \Omega \cap P) = ind(T + my_0, \Omega \cap P)$$

and thus it follows that $ind(T, \Omega \cap P) = 0$.

Corollary 7.3.11. Let X be a real Banach space, Ω_1, Ω_2 be two open bounded subsets of X and P be a cone in X. If $0 \in \Omega_1 \subset \Omega_2$, $T : \overline{\Omega_2} \cap P \to P$ is a continuous countably condensing mapping and the following conditions are satisfied:

(1) $Tx \not\geq x$ for all $x \in \partial\Omega_1 \cap P$;

(2) $Tx \not\leq x$ for all $x \in \partial\Omega_2 \cap P$;

then T has a fixed point in $\overline{\Omega_2 \setminus \Omega_1} \cap P$.

Proof. We may assume that $x \neq Tx$ for all $x \in \partial\Omega_i \cap P$, $i = 1, 2$. By the assumption (1) and the homotopy property (4) of Theorem 7.2.2, we have

$$ind(T, \Omega_1 \cap P) = 1.$$

On the other hand, for any $y_0 \in P$ with $y_0 \neq 0$, we know, by the assumption (2), that $x \neq Tx + \lambda y_0$ for all $x \in \partial\Omega_2 \cap P$ and $\lambda \geq 0$. Therefore, it follows from Theorem 7.3.10 that $ind(T, \Omega_2 \cap P) = 0$ and so

$$ind(T, (\Omega_2 \setminus \overline{\Omega_1}) \cap P) = -1.$$

Therefore, T has a fixed point in $(\Omega_2 \setminus \overline{\Omega_1}) \cap P$. This completes the proof.

Corollary 7.3.12. Let X be a real Banach space, Ω_1, Ω_2 be two open bounded subsets of X and P be a quasinormal cone in X. If $0 \in \Omega_1 \subset \Omega_2$, $T : \overline{\Omega_2} \cap P \to P$ is a continuous countably k-set contraction with $k < \sigma$, where σ is the quasinormality constant of P and the following conditions are satisfied:

(1) $\|Tx\| \leq \|x\|$ for all $x \in \partial\Omega_1 \cap P$;

(2) $\|Tx\| \geq \|x\|$ for all $x \in \partial\Omega_2 \cap P$;

then T has a fixed point in $\overline{\Omega_2 \setminus \Omega_1} \cap P$.

Proof. We may assume that $x \neq Tx$ for all $x \in \partial\Omega_i \cap P$, $i = 1, 2$. By the assumption (1) and the homotopy property (4) of Theorem 7.2.2, we have

$$ind(T, \Omega_1 \cap P) = 1.$$

On the other hand, we have, by the assumption (2) and Corollary 7.3.9, that $ind(T, \Omega_2 \cap P) = 0$ and so

$$ind(T, (\Omega_2 \setminus \overline{\Omega_1}) \cap P) = -1.$$

Therefore, T has a fixed point in $(\Omega_2 \setminus \overline{\Omega_1}) \cap P$. This completes the proof.

Theorem 7.3.13. Let X be a real Banach space, Ω be an open bounded subset of X and P be a cone in X. If $T : \overline{\Omega} \cap P \to P$ is a continuous countably condensing mapping, $S : \partial\Omega \cap P \to P$ is a continuous compact mapping and the following conditions are satisfied:

(1) $\inf\{\|Sx\| : x \in \partial\Omega \cap P\} > 0$;

(2) $x \neq Tx + \lambda Sx$ for all $x \in \partial\Omega \cap P$ and $\lambda > 0$;

then $ind(T, \Omega \cap P) = 0$.

Proof. By the assumption (1), we know that $0 \notin \overline{S(\partial\Omega \cap P)}$. and thus, it follows from the well-known argument, (see Lemma 7.3.2) that

$$0 \notin Conv(\overline{S(\partial\Omega \cap P)}).$$

By Dugundji's extension theorem, there exists a continuous mapping $S^* : \overline{\Omega \cap P} \to Conv(\overline{S(\partial\Omega \cap P)})$ such that $S^*x = Sx$ for all $x \in \partial\Omega \cap P$ and so S^* is compact.

From assumption (2) and the homotopy property (4) of Theorem 7.2.2, we have

$$ind(T, \Omega \cap P) = ind(T + mS^*, \Omega \cap P) \quad \text{for all } m > 0$$

and thus $ind(T, \Omega \cap P) = 0$, for, otherwise, $x = Tx + mS^*x$ has a solution in $\Omega \cap P$ for any $m > 0$, which is a contradiction (note that Ω is bounded and $\inf_{x \in \overline{\Omega} \cap P} \|S^*x\| = a > 0$). This completes the proof.

Theorem 7.3.14. Let X be a real Banach space, Ω_1, Ω_2 be two open bounded subsets of X such that $0 \in \Omega_1 \subset \Omega_2$ and P is a cone in X. If $T : \overline{\Omega_2} \cap P \to P$ is a continuous countably condensing mapping, $S : \partial\Omega_2 \cap P \to P$ is a continuous compact mapping and the following conditions are satisfied:

(1) $\inf\{\|Sx\| : x \in \partial\Omega_2 \cap P\} > 0$;

(2) $Tx \neq \alpha x$ for all $x \in \partial\Omega_1 \cap P$, $\alpha \geq 1$;

then one of the following conclusion holds:

(I) T has a fixed point in $\overline{(\Omega_2 \setminus \overline{\Omega_1})} \cap P$;

(II) $x = Tx + \lambda Sx$ has a solution in $\partial\Omega_2 \cap P$ for some $\lambda > 0$.

Proof. Assume that (II) is not true. Then, by Theorem 7.3.13, we have $ind(T, \Omega_2 \cap P) = 0$. On the other hand, by the assumption (2), we have $ind(T, \Omega_1 \cap P) = 1$ and thus,

$$ind(T, (\Omega_2 \setminus \overline{\Omega_1}) \cap P) = -1.$$

Therefore, T has a fixed point in $(\Omega_2 \setminus \overline{\Omega_1}) \cap P$.

Theorem 7.3.15. Let X be a real Banach space, Ω_i, $i = 1, 2, 3$, be open bounded subsets of X such that $0 \in \Omega_1$, $\overline{\Omega_i} \subset \Omega_{i+1}$, $i = 1, 2$, and P is a cone in X. If $T : \overline{\Omega_3} \cap P \to P$ is a continuous countably condensing mapping satisfying the following conditions:

(1) $\|Tx\| \leq \|x\|$, for all $x \in \partial\Omega_1 \cap P$;

(2) $\inf\left\{\frac{\|Tx\|}{\|x\|} : x \in \partial\Omega_2 \cap P\right\} > \frac{1}{\sigma}$, where σ is the quasinormality constant of P;

(3) $\|Tx\| \leq \|x\|$, for all $x \in \partial\Omega_3 \cap P$.

Then T has two positive fixed points in $(\overline{\Omega_3} \setminus \Omega_1) \cap P$.

Proof. We first prove that T has a fixed point in $(\overline{\Omega_2} \setminus \Omega_1)) \cap P$. We may also assume that $Tx \neq x$ for all $x \in \partial\Omega_1 \cap P$. By the assumption (1), we then have $x \neq tTx$ for all $x \in \partial\Omega_1 \cap P$ and $t \in [0,1]$. Thus,

$$ind(T, \Omega_1 \cap P) = 1. \tag{7.3.2}$$

By the assumption (2) and Theorem 7.3.7, we have

$$ind(T, \Omega_2 \cap P) = 0. \tag{7.3.3}$$

From (7.3.2) and (7.3.3), we deduce $ind(T, (\Omega_2 \setminus \overline{\Omega_1}) \cap P) = -1$ and, consequently, T has a fixed point in $(\Omega_2 \setminus \overline{\Omega_1}) \cap P$. To see that T has a fixed point in $(\overline{\Omega_3} \setminus \Omega_2)) \cap P$, we may also assume that $Tx \neq x$ for all $x \in \partial\Omega_3 \cap P$. By the assumption (3), we then have $x \neq tTx$ for all $x \in \partial\Omega_3 \cap P$ and $t \in [0,1]$. Thus,

$$ind(T, \Omega_3 \cap P) = 1. \tag{7.3.4}$$

From (7.3.3) and (7.3.4), we deduce $ind(T, (\Omega_3 \setminus \overline{\Omega_2}) \cap P) = 1$ and, consequently, T has a fixed point in $(\Omega_3 \setminus \overline{\Omega_2}) \cap P$. This completes the proof.

Finally, we give a different version of Theorem 7.3.15 as follows:

Theorem 7.3.16. Let X be a real Banach space, Ω_i, $i = 1, 2, 3$, be open bounded subsets of X such that $0 \in \Omega_1$, $\overline{\Omega_i} \subset \Omega_{i+1}$, $i = 1, 2$, and P is a cone in X. If $T : \overline{\Omega_3} \cap P \to P$ is a continuous countably condensing mapping satisfying the following conditions:

(1) $Tx \not> x$ for all $x \in \partial\Omega_1 \cap P$;

(2) $Tx \not\leq x$ for all $x \in \partial\Omega_2 \cap P$;

(3) $Tx \not> x$ for all $x \in \partial\Omega_3 \cap P$;

then T has two positive fixed points in $(\overline{\Omega_3} \setminus \Omega_1) \cap P$;

7.4 Perturbations of Condensing Mappings

In this section, let X be a real Banach space, P be a quasinormal cone in X, $T : D(T) \subset X \to X$ be a completely continuous or condensing mapping and $A : D(A) \subset X \to 2^X$ be an accretive operator. We consider the existence of solutions to the mapping equation $x \in -Ax + Tx$ and prove some existence results by using the index theory in Section 7.2

Theorem 7.4.1. Let P be a cone in a Banach space E, $A : D(A) \subseteq P \to 2^P$ be an accretive operator with $P = (I + A)(D(A))$, $0 \in \Omega$ be an open bounded subset of E and $K : \overline{\Omega} \cap P \to P$ be a continuous compact mapping. Suppose that

$$Kx \leq x \quad \text{for all } x \in \partial\Omega \cap P.$$

Then $-A + K$ has a fixed point in $\overline{\Omega} \cap D(A)$.

Proof. It is easy to see that $x \in -Ax + Kx$ if and only if $x = (I + A)^{-1}Kx$. If $x \in -Ax + Kx$ for some $x \in \partial\Omega \cap D(A)$, then the conclusion is true and so we may assume that $x \notin -Ax + Kx$ for all $x \in \partial\Omega \cap D(A)$.

Put $H(t, x) = (I + A)^{-1}tKx$ for all $(t, x) \in [0, 1] \times \overline{\Omega} \cap P$. Then we have

$$x \neq (I + A)^{-1}tKx \quad \text{for all } (t, x) \in [0, 1] \times \partial\Omega \cap P.$$

Indeed, if $x = (I + A)^{-1}tKx$ for some $(t, x) \in [0, 1] \times \partial\Omega \cap P$, then $t \neq 1$ and $tKx \in x + Ax$. However, $Ax \geq 0$ and hence we have $tKx \geq x$. Therefore, it follows that $t > 0$ and $Kx \geq t^{-1}x > x$, which is a contradiction. Hence we have

$$ind(I + A)^{-1}K, \Omega \cap P) = deg(0, \Omega \cap P) = 1.$$

Therefore, $-A + K$ has a fixed point in $\overline{\Omega} \cap D(A)$. This completes the proof.

Theorem 7.4.2. Let P be a cone in a Banach space E, $A : D(A) \subseteq P \to 2^P$ be an accretive operator with $P = (I + A)(D(A))$, $0 \in \Omega$ be an open bounded subset of E and $K : \overline{\Omega} \cap P \to P$ be a continuous compact mapping. Suppose that

$$\|Kx\| \leq \|x\| \quad \text{for all } x \in \partial\Omega \cap P.$$

Then $-A + K$ has a fixed point in $\overline{\Omega} \cap D(A)$.

Proof. We may assume that $x \notin -Ax + Kx$ for all $x \in \partial\Omega \cap D(A)$. Let $H(t, x)$ be defined as in Theorem 7.4.1. Then we have

$$x \neq H(t, x) \quad \text{for all } (t, x) \in [0, 1] \times \partial\Omega \cap P.$$

Indeed, if $x = (I + A)^{-1}tKx$ for some $(t, x) \in [0, 1] \times \partial\Omega \cap P$, then we have

$$\|x\| \leq \|(I + A)^{-1}tKx\| \leq \|tKx\|$$

since $(I + A)^{-1}0 = 0$ and $(I + A)^{-1}$ is nonexpansive. Therefore $t = 1$, which contradicts our assumption. Thus, we have

$$ind(I + A)^{-1}K, \Omega \cap P) = ind(0, , \Omega \cap P) = 1$$

and so T has a fixed point. This completes the proof.

Theorem 7.4.3. Let X be a real Banach space, P be a quasinormal cone in X, $A : D(A) \subseteq P \to 2^P$ be an accretive mapping with $P = (I + \lambda A)P$ for all $\lambda > 0$ and Ω_i, $i = 1, 2$, be two open bounded subsets of X with $0 \in \Omega_1 \subset \Omega_2$. Let $T : \overline{\Omega_2} \cap P \to P$ be a continuous countably condensing mapping. Suppose that the following conditions are satisfied:

(1) $\|Tx\| \leq \|x\|$ for all $x \in \partial\Omega_1 \cap P$;

(2) $\inf\left\{\frac{\|Tx\|}{\|x\|+\|v\|} : x \in \partial\Omega \cap P,\, v \in Ax\right\} > \frac{1}{\sigma}$; where σ is the quasinormality constant of P.

Then $-A + T$ has a fixed point in $\overline{\Omega_2 \setminus \Omega_1} \cap P$.

Proof. We may assume that $x \notin -Ax + Tx$ for all $x \in \partial\Omega_i \cap P$ for $i = 1, 2$. Set

$$U_1 = (I + A)(\Omega_1 \cap P), \quad U_2 = (I + A)(\Omega_2 \cap P).$$

Since $(I + A)^{-1}$ is nonexpansive, it follows that U_1, U_2 are open subsets of P and $T(I + A)^{-1}$ is countably condensing. It is easy to see that $-A + T$ has a fixed point in $\overline{\Omega_2 \setminus \Omega_1} \cap P$ if and only if $T(I + A)^{-1}$ has a fixed point in $\overline{U_2 \setminus U_1}$ and the set of fixed points of $T(I + A)^{-1}$ are bounded in $\overline{U_2 \setminus U_1}$, and so we may simply assume that U_1 and U_2 are bounded. It is obvious that $H(\cdot, \cdot) : [0, 1] \times \overline{\Omega_1} \cap P \to P$ defined by

$$H(t, x) = tT(I + A)^{-1}x \quad \text{for all } (t, x) \in [0, 1] \times \overline{\Omega_1} \cap P$$

is a homotopy of countably condensing mappings.

Now, we claim that $x \neq H(t, x)$ for all $(t, x) \in [0, 1] \times \partial U_1$. In fact, if not, then there exist $t_0 \in [0, 1]$ and $x_0 \in \partial U_1$ such that $x_0 = t_0 T(I + A)^{-1}x_0$. Put $z_0 = (I + A)^{-1}x_0$, then $z_0 \in \partial\Omega_1 \cap P$. Take $v_0 \in Az_0$ such that $x_0 = z_0 + v_0$ and then we have

$$\|z_0\| \leq \|z_0 + v_0\| = \|x_0\| = t_0\|Tz_0\|.$$

From the assumption (1), we must have $t_0 = 1$, which contradicts the fact that $-A + T$ has no fixed point on $\partial\Omega_1 \cap P$. Thus we have

$$ind(T(I + A)^{-1}, U_1) = ind(0, U_1) = 1$$

since $0 \in A0 \subset U_1$. By the definition of quasinormality constant of P and the assumption (2), we may take $\epsilon > 0$ and $y_0 \neq 0 \in P$ such that $\|x + y_0\| \geq (\sigma - \epsilon)\|x\|$ for all $x \in P$ and

$$\inf\left\{\frac{\|Tx\|}{\|x\| + \|v\|} : x \in \partial\Omega \cap P,\, v \in Ax\right\} > \frac{1}{\sigma - \epsilon}. \tag{7.4.1}$$

Next, we claim that $x \neq T(I + A)^{-1}x + \lambda y_0$ for all $x \in \partial U_2$ and $\lambda \geq 0$. If not, then there exist $x_0 \in \partial U_2$ and $\lambda_0 > 0$ such that $x_0 = T(I+A)^{-1}x_0 + \lambda y_0$. Put $z_0 = (I + A)^{-1}x_0$, and then $z_0 \in \partial\Omega_2 \cap P$. There exists $v_0 \in Az_0$ such that $x_0 = z_0 + v_0$, and so we have

$$\|z_0 + v_0\| = \|x_0\| = \|Tz_0 + \lambda y_0\| \geq (\sigma - \epsilon)\|Tz_0\|,$$

which contradicts (7.4.1). By the homotopy property (4) of Theorem 7.2.2, we get

$$ind(T(I+A)^{-1}, U_2) = ind(T(I+A)^{-1} + \lambda y_0) \quad \text{for all } \lambda > 0$$

and so $ind(T(I+A)^{-1}, U_2) = 0$. Therefore, we have

$$ind(T(I+A)^{-1}, U_2 \setminus \overline{U_1}) = -1$$

and so $T(I+A)^{-1}$ has a fixed point in $U_2 \setminus \overline{U_1}$, i.e., $-A+T$ has a fixed point in $(\Omega_2 \setminus \overline{\Omega_1}) \cap P$. This completes the proof.

Remark. Nonexpansive mappings with dissipative perturbations in cones have been studied by Chang et al. [53].

7.5 Index Theory for Nonself Mappings

In this section, first, let $K \subset R^n$ be a nonempty cone, $\Omega \subset R^n$ be a nonempty bounded subset with $\Omega \cap K \neq \emptyset$ and $f : \overline{\Omega} \cap K \to R^n$ be a continuous mapping. We recall the following definition from [177].

Definition 7.5.1. [182] A mapping $f : \overline{\Omega} \cap K \to R^n$ is said to be generalized inward if $d(x, f(x)) \neq d(f(x), K)$ for all $x \in \overline{\Omega} \cap K$ with $f(x) \notin K$, where $d(f(x), K) = \inf_{y \in K} d(f(x), y)$.

Now, assume that $f : \overline{\Omega} \cap K \to R^n$ is a continuous generalized inward mapping. Let $r : R^n \to K$ be a metric projection, i.e., $d(x, r(x)) = d(x, K)$ for all $x \in R^n$. Such a mapping always exists and is unique. We know that $rf : \overline{\Omega} \cap K \to K$ is continuous and $rf(x) \neq x$ for all $x \in \partial\Omega \cap K$. Otherwise, if $rf(x) = x$ for some $x \in \partial\Omega \cap K$, then $d(x, f(x)) = d(rf(x), f(x)) = d(f(x), K)$, which is a contradiction.

Now, we define

$$ind(f, \Omega \cap K) = ind(rf, \Omega \cap K), \tag{7.5.1}$$

where $ind(rf, \Omega \cap K)$ is the fixed point index in Section 7.2, and this is called the fixed point index for generalized inward mapping.

Theorem 7.5.1. The fixed point index for generalized inward mapping has the following properties:

(1) $ind(x_0, \Omega \cap K) = 1$ if $x_0 \in \Omega \cap K$;

(2) If $ind(f, \Omega \cap K) \neq 0$, then $f(x) = x$ has a solution in $\Omega \cap K$;

(3) If Ω_1, Ω_2 are two open bounded disjoint subsets, then $ind(f, (\Omega_1 \cup \Omega_2) \cap K) = ind(f, \Omega_1 \cap K) + ind(f, \Omega_2 \cap K)$;

(4) Let $H(t,x) : [0,1] \times \overline{\Omega} \cap K \to R^n$ be a continuous mapping satisfying $H(t,x) \neq x$ for all $(t,x) \in [0,1] \times \partial\Omega \cap K$ and $d(f(x), H(t,x)) \neq d(f(x), x)$ for all $(t,x) \in \overline{\Omega} \cap K$ with $H(t,x) \notin K$. Then $ind(H(t,\cdot), \Omega \cap K)$ does not depend on $t \in [0,1]$.

Now, we assume that E is a real Banach space, $\Omega \subset E$ is open and bounded and $P \subset E$ is a cone with $\overline{\Omega} \cap P \neq \emptyset$. For all $x \in P$, $I_P(x) = \{x + \lambda(y - x) : \lambda \geq 1, y \in P\}$ is called the inward set of x relative to P.

Definition 7.5.2. Let $T : \overline{\Omega} \cap P \to E$ be a mapping.

(1) If $d(Tx, P) \neq d(x, Tx)$ for all $Tx \notin P$, then T is said to be a generalized inward mapping.

(2) If $Tx \in I_P(x)$ for all $x \in \overline{\Omega} \cap P$, then T is said to an inward mapping.

(3) If $Tx \in \overline{I_P(x)}$ for each $x \in \overline{\Omega} \cap P$, then T is said to a weakly inward mapping.

It is obvious that an inward mapping is weakly inward.

Proposition 7.5.3. If $T : \overline{\Omega} \cap P \to E$ is weakly inward, then T is generalized inward.

Proof. For all $x \in \overline{\Omega} \cap P$ with $Tx \notin P$, we have $d(Tx, I_P(x)) = 0$. Thus, there exists $y \in I_P(x)$ such that $d(Tx, y) < d(x, Tx)$. On the other hand, there exists $z \in P$ such that $z = ty + (1-t)x$ for some $t \in (0,1)$. Thus, we have

$$\begin{aligned} d(Tx, P) &< d(Tx, z) = d(Tx, ty + (1-t)x) \\ &\leq td(Tx, y) + (1-t)d(x, Tx) = d(x, Tx). \end{aligned}$$

Therefore, T is generalized inward on $\overline{\Omega} \cap P$. This completes the proof.

In the following, we assume that there exists a continuous metric projection $r : E \to P$. Assume that $Tx \neq x$ for all $x \in \partial\Omega \cap P$ and then we have $rTx \neq x$ for all $x \in \partial\Omega \cap P$. Otherwise, we have $rTx = x$ for some $x \in \partial\Omega \cap P$. Then $d(x, Tx) = d(rTx, Tx) = d(Tx, P)$, which is a contradiction. Thus, $ind(rT, \Omega \cap P)$ is well defined. Now, we define

$$ind(T, \Omega \cap P) = ind(rT, \Omega \cap P). \tag{7.5.2}$$

If $r_1, r_2 : E \to P$ are two continuous metric projections, then $\{(tr_1 + (1-t)r_2\}_{t \in [0,1]}$ is a family of continuous metric projections. One can easily see that $(tr_1 + (1-t)r_2)Tx \neq x$ for all $x \in \partial\Omega \cap P$, thus $ind([tr_1 + (1-t)r_2]T, \Omega \cap P)$ does not depend on $t \in [0,1]$ by Theorem 7.2.3. Therefore, $ind(rT, \Omega \cap P)$ does not depend on r, and so $ind(T, \Omega \cap P)$ is well defined.

Theorem 7.5.4. The fixed point index defined by (7.5.2) has the following properties:

(1) $ind(x_0, \Omega \cap P) = 1$ if $x_0 \in \Omega \cap P$;

(2) If $ind(T, \Omega \cap K) \neq 0$, then $Tx = x$ has a solution in $\Omega \cap K$;

(3) If Ω_1, Ω_2 are two open bounded disjoint subsets, then $ind(T, (\Omega_1 \cup \Omega_2) \cap P) = ind(T, \Omega_1 \cap P) + ind(T, \Omega_2 \cap P)$;

(4) Let $H(t, x) : [0, 1] \times \overline{\Omega} \cap P \to E$ be a continuous compact mapping satisfying $H(t, x) \neq x$ for all $(t, x) \in [0, 1] \times \partial\Omega \cap P$ and $d(x, H(t, x)) \neq d(H(t, x), P)$ for all $(t, x) \in \overline{\Omega} \cap P$ with $H(t, x) \notin P$. Then $ind(H(t, \cdot), \Omega \cap P)$ does not depend on $t \in [0, 1]$.

Remark. For more details about the results in this section, we refer the reader to [182].

7.6 Applications to Integral and Differential Equations

In this section, we give some applications to the integral and differential equations by using the results of the previous sections.

Example 7.6.1. Let $K(t, s) : [a, b] \times [a, b] \to [0, +\infty)$ be a continous function and $f(t, x) : [a, b) \times R \to [0, +\infty)$ be a continuous function. Suppose the following conditions are satisfied:

(1) For each $t \in [a, b]$, $f(t, x)$ is increasing in x;

(2) $\int_a^b f(s, c)ds < cM^{-1}$, where $M = \max\{K(t, s) : (t, s) \in [a, b] \times [a, b]\}$ and $c > 0$ is a constant;

(3) $f(t, s) \geq \alpha s^\gamma$ for all $s \in [0, \epsilon_0)$, where $\epsilon_0 > 0, \alpha > 0$ and $0 < \gamma < 1$ are constants;

(4) $K(t, s) \not\equiv 0$ for all $(t, s) \in [a, b] \times [a, b]$.

Then the integral equation:

$$x(t) = \int_a^b K(t, s)f(t, x(s))ds \qquad (E\ 7.6.1)$$

has a nontrivial non-negative solution in $C([a, b])$.

Proof. Define a mapping $T : C([a, b]) \to C([a, b])$ by

$$Tx(t) = \int_a^b K(t, s)f(t, x(s))ds \quad \text{for all } x(\cdot) \in C([a, b]). \qquad (7.6.1)$$

Obviously, T is continuous and compact and finding a solution of (E 7.6.1) is equivalent to finding a fixed point of T. Put

$$P = \{x(\cdot) \in C([a,b]) : x(t) \geq 0, \, t \in [a,b]\}.$$

Then P is a cone in $C([a,b])$. By the assumption (1) and (2), we have

$$\|Tx(\cdot)\| < c, \|x(\cdot)\| = c, x(\cdot) \in P. \tag{7.6.2}$$

By the assumption (4), there exist $(t_0, s_0) \in [a,b] \times [a,b], \delta > 0$ and $\beta > 0$ such that $K(t,s) \geq \beta$ for all $(t,s) \in [t_0 - \delta, t_0 + \delta] \times [s_0 - \delta, s_0 + \delta]$. Thus we have

$$\|Tx(\cdot)\| \geq \int_{s_0-\delta}^{s_0+\delta} \beta f(s, x(s))ds.$$

By the assumption (3), if $\|x(\cdot)\| < \min\{1, \epsilon_0\}$, we have

$$\|Tx(\cdot)\| \geq 2\beta\delta\|x(\cdot)\|^\gamma. \tag{7.6.3}$$

Therefore, we have $\|Tx(\cdot)\| > \|x(\cdot)\|$ for $\|x(\cdot)\| = r$ and $x(\cdot) \in P$ with r sufficiently small. We take $\Omega = B(0,c)$ and $\Omega_0 = B(0,r)$. By Corollary 7.3.4, we know that T has a fixed point in $\overline{\Omega \setminus \Omega_0} \cap P$, i.e., ($E$ 7.6.1) has a nontrivial non-negative solution.

Example 7.6.2. Consider the boundary value problem:

$$\begin{cases} x''(t) + f(t, x(t)) = 0, & t \in [0,1], \\ x(0) = x(1), \ x'(0) = -x'(1). \end{cases} \tag{E 7.6.2}$$

Assume that f is continuous and satisfies the following conditions:

(1) For all $t \in [0,1]$, $f(t,x)$ is increasing in x;

(2) $\int_0^1 f(s,c)ds < \frac{3c}{4}$;

(3) $f(t,s) \geq \alpha s^\gamma$ for all $s \in [0, \epsilon_0)$, where $\epsilon_0 > 0$, $\alpha > 0$ and $0 < \gamma < 1$ are constants.

Then (E 7.6.2) has a nontrivial non-negative C^2 solution.

Proof. It is well known that (E 7.6.2) is equivalent to the following integral equation:

$$x(t) = \int_0^1 G(t,s) f(s, x(s))ds, \tag{E 7.6.3}$$

where $G(t,s)$ is the Green function defined by

$$G(t,s) = \begin{cases} \frac{1}{3}(t+1)(2-s), & t \leq s, \\ \frac{1}{3}(s+1)(2-t), & t > s. \end{cases}$$

One may easily see that $M = \max\{G(t,s) : (t,s) \in [0,1] \times [0,1]\} < \frac{4}{3}$ and f, G satisfy the conditions of Example 7.6.1. Therefore, (E 7.6.3) has a nontrivial non-negative solution in $C([0,1])$, i.e., (E 7.6.1) has a nontrivial non-negative C^2 solution.

7.7 Exercises

1. Let $\Omega \subset R^n$ be measurable subset with $m(\Omega) < +\infty$, $1 \leq p \leq \infty$ and $P \subset L^p(\Omega)$ be given by $P = \{f(\cdot) \in L^p(\Omega) : f(x) \geq 0$, almost all $x \in \Omega\}$. Show that P is fully regular.

2. Let $P \subset c_0$ be given by $P = \{(x_i) \in c_0 : x_i \geq 0,\ i = 1, 2, \cdots\}$. Show that P is regular, but not fully regular.

3. Let E be a reflexive Banach space and $P \subset E$ be a cone. Prove that the following conclusions are equivalent:

   (a) P is normal;

   (b) P is regular;

   (c) P is fully regular.

4. Let E be a normed space and $P \subset E$ be a cone. If P has the nonempty interior, then show that P is reproducing.

5. Let E be a normed space, $P \subset E$ be a cone and $P^* \subset E^*$ be defined by $P^* = \{f \in E^* : f(x) \geq 0$ for all $x \in P\}$. If P is reproducing, then show that P^* is a cone in E^*.

6. Let $P \subset L^1(\Omega)$ be defined by $P = \{f(\cdot) \in L^1(\Omega) : f(x) \geq 0$, almost all $x \in \Omega\}$, where $m(\Omega) < \infty$. Show that P allows plastering.

7. Let E be a Banach space, $P \subset$ be a normal cone, $[a, b] = \{x \in E : a \leq x \leq b\}$ and $T : [a, b] \to [a, b]$ be a continuous condensing mapping satisfying $Ax \leq Ay$ for all $x, y \in [a, b]$ with $x \leq y$. Show that T has a fixed point in $[a, b]$.

8. Let E be a locally convex space, $P \subset E$ be a cone, $\Omega \subset E$ be an open subset with $\Omega \cap P \neq \emptyset$ and $T : \overline{\Omega} \cap P \to P$ be a continuous mapping such that $T(\overline{\Omega} \cap P)$ is relatively compact in E. Assume that $x \neq Tx$ for all $x \in \partial\Omega \cap P$. Construct the fixed point index theory for T on $\Omega \cap P$.

9. Let E be a locally convex space, $P \subset E$ be a cone, $\Omega \subset E$ be an open subset with $0 \in \Omega$ and $\Omega \cap P \neq \emptyset$ and $T : \overline{\Omega} \cap P \to P$ be a continuous mapping such that $T(\overline{\Omega} \cap P)$ is relatively compact in E. Assume that $x \neq tTx$ for all $x \in \partial\Omega \cap P$ and $t \in [0, 1)$. Show that T has a fixed point in $\overline{\Omega} \cap P$.

10. Let E be a Banach space, $P \subset E$ be a cone and $A : P \to E$ be a continuous accretive operator. Assume that, for all $x \in P$, there exists $\alpha(x) > 0$ such that $Ax \leq \alpha(x)x$. Show that $(\lambda I + A)(P) \supseteq P$ for all $\lambda > 0$.

11. Let $a(s,t) : [0,1] \times [0,1] \to [0,+\infty)$ be a continuous function satisfying the following conditions:

    (1) $[a(s,t) - a(s,r)](t-r) \geq 0$;

    (2) $a(s,t) \leq ct$ for some constant $c > 0$ and all $(s,t) \in [0,1] \times [0,1]$. Let $k(s,t) : [0,1] \times [0,1] \to [0,+\infty)$ be a continuous function such that $k(s,t) \leq \alpha t + \beta$ for all $(s,t) \in [0,1] \times [0,1]$, where $\alpha \in (0,1)$, $\beta > 0$ are constant. Show that the following integral inequation:

    $$x(t) + a(t, x(t)) - \int_0^1 k(t, x(s))ds = 0 \quad \text{for all } t \in [0,1]$$

    has a solution $x(\cdot) \in C([0,1])$.

12. Let a, b, c, d be non-negative numbers such that $e = ac + bc + ad > 0$ and $f(x,y) : [0,1] \times R \to R$ be defined by $f(x,y) = \Sigma_{i=1}^m a_i(x) y^{\alpha_i}$ for all $(x,y) \in [0,1] \times R$, where $a_i(t) : [0,1] \to [0,+\infty)$ is continuous for $i = 1, 2 \cdots, m$ and $\alpha_i > 0$ for $i = 1, 2 \cdots, m$. Suppose that there exist $1 \leq j, k \leq m$ such that $\alpha_j(t) < 1$, $\alpha_k > 1$, $a_j(t) a_k(t) > 0$ for all $t \in [0,1]$ and $\Sigma_{i=1}^m \int_0^1 a_i(t)dt < f^{-1}$, where $f = e^{-1}(4ac)^{-1}$ if $ac \neq 0$, $f = e^{-1}(bc + bd)$ if $a = 0$ and $f = e^{-1}(ad + bd)$ if $c = 0$. Show that the following equation:

    $$\begin{cases} x''(t) = -f(t, x(t)), & t \in [0,1], \\ ax(0) - bx'(0) = 0, & cx(1) + dx'(1) = 0 \end{cases}$$

    has two non-negative nontrivial solutions in $C^2([0,1])$ by using the Green function $G(t,s)$ defined as

    $$G(t,s) = \begin{cases} e^{-1}(at + b)[c(1-s) + d], & t \leq s, \\ e^{-1}(as + b)[(c(1-t) + d], & t > s. \end{cases}$$

Chapter 8

REFERENCES

1. A. Addou and B. Mermri, Topological degree and application to a parabolic variational inequality problem, Inter. J. Math. Sci. 25(2001) 273–287.

2. R. P. Agarwal, M. Meehan and D. O'Regan, Fixed Point Theory and Application, Cambridge University Press, New York, 2001.

3. R. Agarwal and D. O'Regan, An index theory for P-concentrative J-maps, Appl. Math. Lett. 16(2003), 1265–1271.

4. S. Aizicovici and Y. Q. Chen, Note on the topological degree of the subdifferential of a lower semicontinuous convex function, Proc. Amer. Math. Soc. 126(1998), 2905–2908.

5. H. Alex, S. Hahn and L. Kaniok, The fixed point index for noncompact mappings in non-locally convex topological vector spaces, Comment. Math. Univ. Carolinae 32(1994), 249–257.

6. M. Altman, A fixed theorem for completely continuous operators in Banach spaces, Bull. Acad. Polon. Sci. Sér. Sci. Math. Astonom. Phys. 3(1955), 490–513.

7. H. Amann, On the number of solutions of nonlinear equations in Banach spaces, J. Funct. Anal. 11(1972), 346–384.

8. H. Amann, Fixed point equations and nonlinear eigenvalue problems in ordered Banach spaces, SIAM Review 18(1976), 620–709.

9. H. Amann, A note on degree theory for gradient mappings, Proc. Amer. Math. Soc. 85(1982), 591–595.

10. H. Amann and S. Weiss, On the uniqueness of the topological degree, Math. Z. 130(1973), 39–54.

11. J. Appell, M. Väth and A. Vignoli, Compactness and existence results for ordinary differential equations in Banach spaces, Z. Anal. Anw. 18(1988), 469–484.

12. J. P. Aubin and A. Cellina, Differential Inclusions, Springer-Verlag, Berlin, 1984.

13. J. P. Aubin and I. Ekeland, Applied Nonlinear Analysis, Wiley, New York, 1984.

14. J. P. Aubin and H. Frankowska, Set-Valued Analysis, Birkhauser, Boston, 1990.

15. R. Bader, A topological fixed-point index theory for evolution inclusions, Z. Anal. Anw. 20 (2001) 3–16.

16. S. Banach, Theory of linear operations (English translation), North-Holland, 1987.

17. V. Barbu, Nonlinear Semigroups and Differential Equations in Banach Spaces, Noordhoff, Leyden, 1976.

18. R. I. Becker, Periodic solutions of semilinear equations of evolution of compact type, J. Math. Anal. Appl. 82(1982), 33–48.

19. V. Benci, A geometrical index for the group S^1 and some applications to the study of periodic solutions of ordinary differential equations, Comm. Pure Appl. Math. 34(1981), 393–432.

20. J. Berkovits and V. Mustonen, On the degree for mappings of monotone type, Nonlinear Anal. 12(1986), 1373–1383.

21. J. Berkovits and V. Mustonen, Topological degree for perturbation of linear maximal monotone mappings and applications to a class of parabolic problems, Rend. Mat. VII(1992), 597–621.

22. U. G. Borisovic, B. D. Gelman, A. D. Myskis and V. V. Obuhowskii, Topological methods in fixed point theory of multivalued mappings, Uspiehi Mat. Nauk. 1(1980), 59–126.

23. K. Borsuk, Drei Sätze über die n-dimensional Euklidische Sphäre, Fund. Math. 20(1933), 177–190.

24. D. Bothe, Upper semicontinuous perturbations of m-accretive operators and differential inclusions with dissipative right-hand side, Proc. Workshop "Topological Methods in Differential Inclusions", Warsaw, 1994.

25. J. Bourgain, H. Brézis and P. Mironescu, Lifting, degree, and distributional Jacobian, Comm. Pure and Appl. Math. Vol. 58, 2005, 529–551.

26. H. Brézis, Operateurs maxinaux monotones, North-Holland, 1973.

27. H. Brézis, Degree theory: old and new, in Topological Nonlinear Analysis II: Degree, Singularity and Variations (M. Matzeu and A. Vignoli, Ed.) Birkhauser, 1997, 87–108.

28. H. Brézis, Y. Li, P. Mironescu and L. Nirenberg, Degree and Sobolev spaces, Top. Meth. Nonlinear Anal. 13(1999), 181–190.

29. H. Brézis, L. Nirenberg, Degree theory and BMO, Part I: compact manifolds without boundaries, Selecta Math. 1(1995), 197–263.

30. H. Brézis and F. E. Browder, A general ordering principle in nonlinear functional analysis, Advances in Math. 21(1976), 355–364.

31. H. Brézis, M. G. Crandall and A. Pazy, Perturbations of nonlinear maximal monotone sets, Comm. Pure Appl. Math. 23(1970), 123–144.

32. L. E. J. Brouwer, Uber Abbildung der Mannigfaltigkeiten, Math. Ann. 70(1912), 97–115.

33. F. E. Browder, The fixed point theory of multi-valued mappings in topological vector spaces, Math. Ann. 177(1968), 283–301.

34. F. E. Browder, Nonlinear Operators and Nonlinear Equations of Evolution in Banach Spaces, Proc. Symp. Pure Math., Amer. Math. Soc. 18, Part 2, 1976.

35. F. E. Browder, Fixed point theory and nonlinear problems, Bull. Amer. Math. Soc. 1(1983), 1–39.

36. F. E. Browder, Degree theory for nonlinear mappings, Proc. Symp. Pure Math. Soc. 45(1986), 203–226.

37. F. E. Browder and R. D. Nussbaum, The topological degree for noncompact nonlinear mappings in Banach spaces, Bull. Amer. Math. Soc. 74(1968), 671–676.

38. F. E. Browder and W. V. Petryshyn, Approximation methods and the generalized topological degree for nonlinear mappings in Banach spaces, J. Funct. Anal. 3(1969), 217–274.

39. R. F. Brown, An Introduction to Nonlinear Analysis, Birkhauser, 1993.

40. A. Buica, Contributions to the coincidence degree theory of homogenous operators, Pure Math. Appl. 11(2000), 153–159.

41. A. Buica and Llibre, Averaging methods for finding periodic orbits via Brouwer degree, Bull. Sci. Math. 128(2004), 7–22.

42. G. L. Cain and M. Z. Nashed, Fixed points and stability for a sum of two operators in locally convex spaces, Pacific J. Math. 39 (1971), 581–592.

43. A. Capietto, M. Henrard, J. Mawhin and F. Zanolin, A continuation approach to forced superlinear Sturm-Liouville BVP's, Topological Methods in Nonlinear Anal. 3(1994), 81–100.

44. A. Capietto, J. Mawhin and F. Zanolin, A continuation approach to superlinear periodic BVP's, J. Diff. Equat. 88(1990), 347–395.

45. A. Capietto, J. Mawhin and F. Zanolin, Continuation theorems for periodic perturbations of autonomous systems, Trans. Amer. Math. Soc. 329(1992), 41–72.

46. R. Cauty, Solution du probléme de point fixe de Schauder, Fund. Math. 170 (2001), 231–246.

47. A. Cellina and A. Lasota, A new approach to the definition of topological degree for multivalued mappings, Atti. Acad. Naz. Lincei Rend. Cl. Sci. Fis. Mat. Natur. 47(1969), 434–440.

48. K. C. Chang, Variational methods for non-differentiable functionals and their applications to partial differential equations, J. Math. Anal. Appl. 80(1981), 102–129.

49. K. C. Chang, Critical Point Theory and Its Applications, Shanghai Science and Technology Publishing House, Shanghai, 1986.

50. K. C. Chang, Infinite Dimensional Morse Theory and Multiple Solution Problems, Prog. Non. Different. Equat. Appl. 6, Birkhäuser, Boston, 1993.

51. S. S. Chang and Y. Q. Chen , Degree theory for multivalued (S) type mappings and fixed point theorems, Appl. Math. Mech. 11 (1990), 441–454.

52. S. S. Chang and Y. Q. Chen, Coincidence index methods in studying accretive operator equations, Chinese Annals of Math. 5(1993), 579–583.

53. S. S. Chang, Y. Q. Chen, Y. J. Cho and B. S. Lee, Fixed point theorems for nonexpansive operators with disssipative perturbations in cones, Comment. Math. Univ. Carolinae 39(1998), 49–54.

54. Y. Q. Chen, On accretive operators in cones of Banach spaces, Non. Anal. 27(1996), 1125–1135.

55. Y. Q. Chen, Periodic solutions for evolution equations in Hilbert spaces, Yokohama Math. J. 44 (1997), 43–53.

56. Y. Q. Chen, On the semi-monotone operator theory and applications, J. Math. Anal. Appl. 231(1999), 177–192.

57. Y. Q. Chen and Y. J. Cho, On 1-set contraction perturbations of accretive operators in cones of Banach spaces, J. Math. Anal. Appl. 201(1996), 966–980.

58. Y. Q. Chen, Y. J. Cho and L. Yang, Periodic solutions for nonlinear evolution equations, Dynam. Cont. Disc. Impul. Syst. 9(2002), 581–598.

59. Y. Q. Chen and Y. J. Cho, Anti-periodic solutions for semilinear evolution equations, J. Concrete and Appl. Math. 1(2003), 113–124.

60. Y. Q. Chen and Y. J. Cho, Nonlinear Operator Theory in Abstract Spaces and Applications, Nova Science Publishers, New York, 2004.

61. Y. Q. Chen and Y. J. Cho, Topological Degree Theory for Multi-Valued Mappings of Class $(S_+)_L$, Arch. Math. 84(2005), 325–333.

62. Y. Q. Chen, Y. J. Cho and J. S. Jung, Anti-periodic solutions for evolution equations, Math. Computer Model. 40(2004), 1123–1130.

63. Y. Q. Chen, Y. J. Cho and D. O'Regan, On Positive Fixed Points of Countably Condensing Mappings, Dynamics Cont. Discrete and Impul. Systems 12(2005), 519–527.

64. Y. Q. Chen, Y. J. Cho and D. O'Regan, On Perturbations of Accretive Mappings, Appl. Math. Lett. 18(2005), 775–781.

65. Y. Q. Chen, Y. J. Cho and D. O'Regan, Anti-periodic solutions for evolution equations with mappings in the class (S_+), Math. Nachr. 278(2005), 356–362.

66. Y, Q, Chen and J. K. Kim, Anti-periodic solutions for first order semilnear evolution equations, Non. Funct. Anal. Appl. 8(2003), 49–57.

67. Y, Q, Chen and J. K. Kim, Existence of periodic solutions for first-order evolution equations without coercivity, J. Math. Anal. Appl. 282(2003), 801–815.

68. Y. Q. Chen and D. O'Regan, Coincidence degree theory for mappings of class L-(S_+), (submitted).

69. Y. Q. Chen and D. O'Regan, Generalized degree theory for semilinear operator equations, Glasgow Math. J. 48(2006), 65–73.

70. Y. Q. Chen, X. D. Wang and H. X. Xu, Anti-periodic solutions for semilinear evolution equations, J. Math. Anal. Appl. 273(2002), 627–636.

71. S. N. Chow and J. K. Hale, Methods of Bifurcation Theory, Springer, 1982.

72. F. H. Clarke, Optimization and Nonsmooth Analysis, Wiley, New York, 1983.

73. C. Conley and E. Zehnder, A Morse type index theory for flows and periodic solutions to Hamiltonian systems, Comm. Pure Appl. Math. 37(1984), 207–253.

74. M. G. Crandall and T. Liggett, Generation of semigroups of nonlinear transformations on general Banach spaces, Amer. J. Math. 93(1971), 265–298.

75. M. G. Crandall, H. Ishii and P. L. Lions, User's guide to viscosity solutions of second order partial differential equations, Bull. Amer. Math. Soc. 27(1992), 1–67.

76. J. Cronin, Fixed Points and Topological Degree in Nonlinear Analysis, Math. Surveys, No. 11, Amer. Math. Soc., 1964.

77. S. J. Daher, On a fixed point principle of Sadovskii, Nonlinear Anal. 2(1978), 643–645.

78. S. J. Daher, Fixed point theorems for nonlinear operators with a sequential condition, Nonlinear Anal. 3(1979), 59–63.

79. E. N. Dancer, On the indices of fixed points in cones and applications, J. Math. Anal. Appl. 91(1983), 131–151.

80. E. N. Dancer, A new degree for S^1-invariant gradient mappings and applications, Anal. Nonlinear 2(1985), 329–370.

81. E. N. Dancer, Degree theory on convex sets and applications to bifurcation, Calculus of variations and partial differential equations, Topics on geometrical evolution problems and degree theory, Springer, 185–241, 2000.

82. E. N. Dancer, R. D. Nausbaum and C. A. Stuart, Quasinormal cones in Banach spaces, Nonlinear Anal. 7(1983), 539–553.

83. E. N. Dancer and J. Toland, Degree theory for orbits of prescribed period for flows with a first integral, Proc. London Math. Soc. 60(1990), 549–580.

84. B. Decorogna, Weak Continuity and Weak Lower Semicontinuity of Nonlinear Functionals, Lect. Notes Math. 922, Springer, Berlin, 1982.

85. H. Debrunner and P. Flor, Ein Erweiterungssatz fur monotone Mengen, Arch. Math. 15(1964), 445–447.

86. K. Deimling, Nonlinear Functional Analysis, Springer-Verlag, Berlin, 1985.

87. K. Deimling, Multivalued Differential Equations, W. De Gruyter, Berlin, 1992.

88. J. Diestel W. M. Ruess and W. Schachermayer, Weak compactness in $L^1(\mu, X)$, Proc. Amer. Math. Soc. 118(1993), 447–453.

89. W. Y. Ding, A generalization of the Poincare-Birkhoff theorem, Proc. Amer. Math. Soc. 88(1983), 341–346.

90. S. Domachowski, An application of the topological degree and completely continuous extensions of selectors to nonconvex differential inclusions, Nonlinear Anal. 27(1996), 987–997.

91. Y. H. Du, Fixed points of increasing operators in ordered Banach spaces and applications, Appl. Anal. 38(1990), 1–20.

92. Y. H. Du, A degree theoretic approach to N-species periodic competition systems on the whole R^n, Proc. Roy. Soc. Edinburgh 129(1999), 295–318.

93. J. Dugundji and A. Granas, Fixed Point Theory I, Warszawa, 1982.

94. G. Duvaut and J. L. Lions, Les Inequations en Mecanique et en Physique, Dunod, 1970.

95. Z. Dzedzej, Fixed point index theory for a class of nonacyclic multivalued maps, Dissert. Math. 253(1985), 1–53.

96. S. Eilenberg and D. Montgomery, Fixed point theorems for multivalued transformations, Amer. J. Math. 58(1946), 214–222.

97. D. Eisenbud and H. Levine, The Topological degree of a finite C^∞ map germ, in Structural Stability, the Theory of Catastrophes, and Applications in the Sciences, Lecture Notes in Math. 525, Springer-Verlag.

98. I. Ekland, On the variational principle, J. Math. Anal. Appl. 47(1974), 324–353.

99. I. Ekland and H. Hofer, Periodic solutions with prescribed minmal period for convex autonomous Hamiltonian systems, Invent. Math. 81(1985), 155–188.

100. I. Ekland and J. M. Lasry, On the number of periodic trajectories for a Hamiltonian flow on a convex energy surface, Ann. Math. 112(1980), 283–319.

101. I. Ekland and R. Temam, Convex Analysis and Variational Problems, North-Holand, 1976.

102. K. D. Elworthy and A. J. Tromba, Degree theory on Banach manifolds, Proc. Symp. Pure Math. 18, Part I, 86-94, Amer. Math. Soc., RI, 1970.

103. L. H. Erbe, W. Krawcewicz and J. Wu, Leray-Schauder degree for semilinear Fredholm maps and period BVP's of neutral equations, Nonlinear Anal. 15(1990), 747–764.

104. L. H. Erbe, W. Krawcewicz and J. Wu, A composite coincidence degree with applications to BVP's of neutral equations, Trans. Amer. Math. Soc. 335(1993), 459–478.

105. L. C. Evans, On solving certain nonlinear partial differential equations by accretive operator methods, Israel J. Math. 36(1980), 225–247.

106. E. R. Fadell, Recent results in the fixed point theory of continuous maps, Bull. Amer. Math. Soc. 76(1970), 10–29.

107. E. R. Fadell and S. Husseini, Relative cohomological index theories, Advan. Math. 64(1987), 1–31.

108. E. R. Fadell, S. Husseini and P. H. Rabinowitz, Borsuk-Ulam theorems for arbitrary S^1 actions and applications, Trans. Amer. Math. Soc. 274(1982), 345–360.

109. K. Fan, Fixed-point and minimax theorems in locally convex topological linear spaces, Proc. Nat. Acad. Sci. USA 38(1952), 121–126.

110. K. Fan, A generalization of Tychonoff's fixed point theorem, Math. Ann. 142(1961), 305–310.

111. K. Fan, Sur un theoreme minimax, C. R. Acad. Sci. Paris, Groupe 1 142(1962), 3925–3928.

112. K. Fan, Some properties of convex sets related to fixed point theorems, Math. Ann. 266(1984), 519–537.

113. M. Feckan and R. Kollar, Discontinuous wave equations and a topological degree for some classes of multi-valued mappings, Appl. Math. 44(1999), 15–32.

114. V. V. Fillipov, Basic Topological Structures of Odinary Differential Equations, Kluwer Academic Publishers, 1998.

115. P. M. Fitzpatrick, On the structure of the set of solutions of equations of A-proper mappings, Trans. Amer. Math. Soc. 189(1974), 107–131.

116. M. Furi, M. Martelli and A. Vigonoli, On the solvability of nonlinear operators in normed spaces, Ann. Mat. Pure Appl. 124(1980), 321–343.

117. P. M. Fitzpatrick, A generalized degree for uniform limits of A-proper mappings, J. Math. Anal. Appl. 35(1971), 536–552.

118. P. M. Fitzpatrick, On the structure of the set of solutions of equations involving A-proper mappings, Trans. Amer. Math. Soc. 189(1974), 107–131.

119. P. M. Fitzpatrick, Existence results for equations involving noncompact perturbations of Fredholm mappings with applications to differential equations, J. Math. Anal. Appl. 66(1978), 151–171.

120. P. M. Fitzpatrick and W. V. Petryshyn, Fixed point theorems and the fixed point index for multivalued mappings in cones, J. London Math. Soc. 12(1975), 75–85.

121. P. M. Fitzpatrick and J. Pejsachowicz, Orientation and the Leray-Schauder theory for fully nonlinear elliptic bounday value problems. Mem. Amer. Math. Soc. 101(1993).

122. P. M. Fitzpatrick and J. Pejsachowicz, An extension of the Leray-Schauder degree for fully nonlinear elliptc problems, Proc. Symp. Pure Mathematics 45(1986), 425–438.

123. P. M. Fitzpatrick and J. Pejsachowicz, The fundamental group of the space of linear Fredholm operators and global analysis of semilinear equations, Contemp. Math. 72(1998), 47–87

124. I. Fonseca and W. Gangbo, Degree Theory in Analysis and Applications, Clarendon Press, Oxford, 1995.

125. M. Frigon and D. O'Regan, Fixed points of cone-compressing and cone-extending operators in Frechet spaces, Bull. London Math. Soc. 35(2003), 672–680.

126. S. Fucik, M. Kucera and J. Necas, Ranges of nonlinear asymptotically linear operators, J. Diff. Equat. 17(1975), 375–394.

127. S. Fucik, J. Necas, J. Soucek and V. I. Soucek, Spectral Analysis of Nonlinear Operators, Lect. Notes in Math. 346, Springer-Verlag, 1973.

128. M. Furi, M. Martelli and M. P. Pera, Cobifurcating branches of solutions of nonlinear eigenvalue problems in Banach spaces, Ann. Mat. Pura Appl. 135(1983), 119–131.

129. M. Furi, M. Martelli and M. P. Pera, On the existence of forced oscillations for the spherical pendulum, Boll. dell'Unione Mat. Italiana 4-B(1990), 381390

130. M. Furi, M. Martelli and M. P. Pera, A continuation theorem for periodic solutions of forced motion equations on manifolds and applications to bifurcation theory, Pacific J. Math. 160(1993), 219–244.

131. M Furi, M. Martelli and A. Vigonoli, On the solvability of nonlinear operators in normed spaces, Ann. Mat. Pure Appl. 124(1980), 321–343.

132. E. Getzler, Degree theory for Wiener maps, J. Func. Anal. 68 (1986), 388–403.

133. E. Getler, The degree of the Nicolai map, J. Func. Anal. 74(1987), 121–138.

134. R. E. Gaines and J. Mawhin, Coincidence degree and nonlinear differential equations, Lect. Notes in Math. 568, Springer-Verlag, 1977.

135. D. Gilbarg and N. S. Trudinger, Elliptic Partial Differential Equations of Second Order, Springer-Verlag, Berlin, 1983.

136. I. Glicksberg, A further generalization of the Kakutani fixed-point theorem, with application to Nash equilibrium points, Proc. Amer. Math. Soc. 3(1952), 170–174.

137. L. Gorniewicz, Topological Fixed Point Theory of Multivalued Mappings, Kluwer Academic Publishers, 1999.

138. J. P. Gossez, Opérateutrs monotones non-linéaires dans les espaces de Banach nonréflexifs, J. Math. Anal. Appl. 44(1973), 71–87.

139. A. Granas, The Leray-Schauder index and the fixed point theory for arbitrary ANR's, Bull. Soc. Math. France 100(1972), 209–228.

140. A. Granas, KKM-maps and Their Applications to Nonlinear Problems, The Scottish Book, R. D. Mauldin, Ed., Birkhauser, 1981, 45–61.

141. D. J. Guo, A new fixed point theorem, Acta Math. Sinica 24(1981), 444–450.

142. D. J. Guo, Nonlinear Functional Analysis, Shangdon Science and Technology Publishing Press, Shangdon, 1985.

143. D. J. Guo, Partial Order Method in Nonlinear Analysis, Shangdon Science and Technology Publishing Press, Shangdon, 1997.

144. D. Guo and V. Lakshmikantham, Nonlinear Problems in Abstract Cones, Academic Press, 1988.

145. O. Hadzic, On Kakutani's fixed point theorem in topological vector spaces, Bull. Acad. Polon. Sci. Sér. Sci. Math. 30(1982), 141–144.

146. B. R. Halpern and G. M. Bergman, A fixed point theorem for inward and outward maps, Trans. Amer. Math. Soc. 130(1968), 353–358.

147. F. B. Hang and F. H. Lin, Topology of Sobolev mappings II, Acta Math. 191(2003), 55–107.

148. A. Haraux, Anti-periodic solutions of some nonlinear evolution equations, Manuscripta Math., V.63, 1989, 479–505

149. P. Hartman and G. Stampacchia, On some nonlinear elliptic differential functional equations, Acta Math. 115(1966), 271–310.

150. M. Hazewinkel and M. van de Vel, On almost fixed point theory, Canad. J. Math. 30(1978), 673–699.

151. C. J. Himmelberg, Fixed points of compact multifunctions, J. Math. Anal. Appl. 38(1972), 205–207.

152. H. Hofer, Variational and topological methods in partially ordered Hilbert spaces, Math. Ann. 261(1982), 493–514.

153. H. Hofer, A note on the topological degree at a critical point of mountain-pass-type, Proc. Amer. Math. Soc. 90(1984), 309–315.

154. C. D. Horvath and M. Lassonde, Leray-Schauder spaces, Nonlinear Anal. 45(2001), 923–931.

155. S. Hu and Y. Sun, Fixed point index for weakly inward mappings, J. Math. Anal. Appl. 172(1993), 266–273.

156. A. D. Ioff and V. M. Tihemirov, Theory of Extremal Problems, North-Holland, 1979.

157. G. Isac, 0-epi families of mappings, topological degree and optimization, J. Opt. Theory Appl. 42(1984), 51–75

158. S. Ishikawa, Fixed points and iteration of a nonexpansive mapping in a Banach space, Proc. Amer. Math. Soc. 59(1976), 65–71.

159. I. Ize, I. Massabo and A. Vignoli, Degree theory for equivariant maps, 315(1989), 433–510.

160. L. Janos and M. Martelli, Sequentially condensing maps, Univ. u Novom Sadu Zb. Rad. Priod. Mat. Fak. Ser. Mat. 10(1986), 85–94.

161. J. Jost and M. Struwe, Morse-Conley theory for minimal surface of varing topological type, Invent. Math. 102(1990), 465–499.

162. T. Kaczynski and W. Krawcewicz, Solvability of BVP's for the inclusion $u_{tt} - u_{xx} \in g(t, x, u)$ via the theory of multi-valued A-Proper maps, ZAA 7(1988), 337–346.

163. A. G. Kartsatos and I. V. Skrypnik, Topological degree theories for densely defined mappings involving operators of type (S_+), Advan. in Different. Equat. 4(1999), 413–456.

164. A. G. Kartsatos and I. V. Skrypnik, The index of a critical point for nonlinear elliptic operators with strong coefficient growth, J. Math. Soc. Japan 52(2000), 109–137.

165. A. G. Kartsatos and I. V. Skrypnik, The index of a critical point for densely defined operators of type $(S_+)_L$ in Banach spaces, Trans. Amer. Math. Soc. 354(2001), 1601–1630.

166. A. G. Kartsatos and I. V. Skrypnik, A new topological degree theory for densely defined quasibounded $(\widetilde{S}_+)$-perturbations of multivalued maximal monotone operators in reflexive Banach spaces, Abstr. Appl. Anal. 2005(2005), 121–158.

167. T. Kato, Nonlinear semigroups and evolution equations, J. Math. Soc. Japan 19(1967), 508–520.

168. D. Kinderlehrer and G. Stampacchia, An Introduction to Variational Inequalities and Their Applications, Academic Press, New York, 1980.

169. W. A. Kirk, A fixed point theorem for mappings which do not increase distance, Amer. Math. Monthly 72(1965), 1004–1006.

170. V. Klee, Leray-Schauder theory without local convexity, Math. Ann. 141(1960), 286–296.

171. W. Klingenberg, Riemannian Geometry, Walter de Gruyter, 1982.

172. J. Kobayashi and M. Otani, Topological degree for $(S)_+$ mappings with maximal monotone perturbations and its applications to variational inequalities, Nonlinear Anal. 59(2004), 147–172.

173. Y. Kobayashi, Difference approximation of Cauchy problems for quasi-dissipative operators and generation of nonlinear semigroups, J. Math. Soc. Japan 27(1975), 640–665.

174. Y. Komura, Semigroups of operators in locally convex spaces, J. Funct. Anal. 2(1968), 258–296.

175. J. Korevaar, On a question of Brezis and Nirenberg concerning the degree of circle maps, Selecta Math. 5(1999), 107–122.

176. M. A. Krasnoselskii, Positive Solutions of Operator Equations, Groningen, Noordhoff, 1961.

177. M. A. Krasnoselskii, Topological Methods in the Theory of Nonlinear Integral Equations, Macmillan, New York, 1964.

178. M. A. Krasnoselskii and P. P. Zabreiko, Geometrical Methods of Nonlinear Analysis, Springer-Verlag, Berlin, 1984.

179. K. Kuratowski, C. R. Nardzewski, A general theorem on selections, Bull. Acad. Polon. Soc. 13(1965), 397–403.

180. E. Lami-Dozo, Quasinormality in cones in Hilbert spaces, Acad. Roy. Belg. Bull. CI. Sci. 30(1981), 531–541.

181. K. Q. Lan and J. R. L. Webb, A fixed point index for weakly inward A-proper maps, Nonlinear Anal. 28(1997), 315–325.

182. K. Q. Lan and J. R. L. Webb, A fixed point index for generalized inward mappings of condensing type, Trans. Amer. Math. Soc. 349(1997), 2175–2186.

183. R. Leggett and L. Williams, Multiple positive fixed points of nonlinear operators on ordered Banach spaces, Indiana Univ. Math. J. 28(1979), 673–688.

184. J. Leray, Les problemes nonlineaires, Enseign. Math. 30(1936), 141.

185. J. Leray and J. Schauder, Topologie et equations fonctionnelles, Ann. Sci. Ecole. Norm. Sup. 51(1934), 45–78.

186. G. Z. Li, A new fixed point theorem on semicompact 1-set contraction mappings, Proc. Amer. Math. Soc. 97(1986), 277–280.

187. S. J. Li and D. X. Feng, Degree theory for set-valued maximal monotone operators in Hilbert spaces, Acta Math. Sinica 25(1982), 533–541.

188. Y. Y. Li, Degree theory for second order nonlinear elliptic operators and its applications, Comm. in PDE 14(1989), 1541–1578.

189. T. C. Lim, A fixed point theorem for multivalued nonexpansive mappings in a uniformly convex banach space, Bull. Amer. Math. Soc. 80(1974), 1123–1126.

190. C. S. Lin, Topological degree mean field equations on S^2, Duke Math. J. 104(2000), 501–536

191. J. Lindenstrauss and L. Tzafriri, Classical Banach Spaces I, Springer-Verlag, Berlin, 1977.

192. J. L. Lions and E. Magenes, Non-Homogeneous Boundary Value Problems and Applications, I, II, III, Springer-Verlag, Berlin, 1972, 1973.

193. C. G. Liu and Y. M. Long, Iteration inequalities of the Maslov-type index theory with applications, J. Different. Equat. 165(2000), 355–376.

194. N. G. Lloyd, Degree Theory, Cambridge University Press, Cambridge, 1978.

195. Y. M. Long, A Maslov-type index theory for symplectic paths., Topological Methods in Nonlinear Anal. 10(1997), 47–78.

196. G. Lumer, Semi-inner product spaces, Trans. Amer. Math. Soc. 100(1961), 29–43.

197. T. W. Ma, Topological degree for set-valued compact vector fields in locally convex spaces, Dissertationes Math. 92(1972), 1–43.

198. R. H. Martin, Differential equations on closed subsets of a Banach space, Trans. Amer. Math. Soc. 179(1973), 399–414.

199. R. H. Martin, Nonlinear Operators and Differential Equations in Banach Spaces, Wiley, New York, 1976.

200. I. Massabo and C. A. Stuart, Positive eigenvectors for k-set contractions, Nonlinear Anal. 3(1979), 35–44.

201. J. L. Massera and J. J. Schaffer, Linear Differential Equations and Function Spaces, Academic Press, New York, 1966.

202. R. D. Mauldin, The Scottish book: Mathematical problems from the Scottish Cafe, Birkhauser, 1981.

203. J. Mawhin, Topological Degree Methods in Nonlinear Boundary Value Problems, Amer. Math. Soc. 40, Providence, RI, 1979.

204. J. Mawhin, Degré topologique et solutions périodiques des systémes différentiels non linéaires, Bull. Soc. Roy. Sci. Liége 38(1969), 308–398.

205. J. Mawhin, Equivalence theorems for nonlinear operator equations and coincidence degree theory for some mappings in locally convex topological spaces, J. Diff. Equat. 12(1972), 610–636.

206. J. Mawhin, A futher invariance property of coincidence degree in convex spaces, Ann. Soc. Sci. Bruxelles Sér. I 87(1973), 51–57.

207. J. Mawhin, Leray-Schauder continuation theorems in the absence of a prior bounds, Topological Methods in Nonlinear Anal. 9(1997), 179–200.

208. J. Mawhin, Leray-Schauder degree: a half century of extensions and applications, Topological Methods in Nonlinear Anal. 14(1999), 195–228.

209. J. Mawhin and W. Willem, Critical Point Theory and Hamiltonian Systems, Springer-Verlag, Berlin, 1989.

210. S. Mazur, Uber konvexe in linearen normierten Raumen, Studia Math. 4(1933), 70–84.

211. R. E. Megginson, An Introduction to Banach Space Theory, Springer-Verlag, Berlin, 1998.

212. E. Michael, Continuous selections I, Ann. Math. 63(1956), 361–382.

213. M. Mininni, Conicidence degree and solvability of some nonlinear functional equations in normed spaces: a spectral approach, Nonlinear Anal. 2(1978), 597–607.

214. J. Milnor, Analytic proof of the hairy ball theorem and the Brouwer fixed point theorem, Amer. Math. Monthly 85(1978), 521–524.

215. P. S. Milojevic, A generalization of Leray-Schauder's theorem and surjectivity results for multivalued A-proper and pseudo A-proper mappings, Nonlinear Anal. 1(1977), 263–276.

216. P. S. Milojevic, Fredholm alterative and surjectivity results for multivalued $A-$proper and condensing mappings with applications to nonlinear integral and differential equations, Czech. Math. J. 30(1980), 387–417.

217. P. S. Milojevic, Solvability of operator equations involving nonlinear perturbations of Fredholm mappings with nonnegative index and applications, in Lect. Notes Math., 957, pp. 212–228, Springer-Verlag, 1982.

218. G. J. Minty, Monotone (nonlinear) operators in Hilbert spaces, Duke Math. J. 29(1962), 341–346.

219. G. J. Minty, On the generalization of a direct method of calculus of variations, Bull. Amer. Math. Soc. 73(1967), 315–321.

220. H. Mönch, Boundary value problems for nonlinear ordinary differential equations in Banach spaces, Non. Anal. 4(1980), 985–999.

221. H. Mönch and G. F. von Harten, On the Cauchy problems for ordinary differential equations in Bnach spaces, Arch. Math. 39(1982), 153–160.

222. C. Morales, Remarks on compact perturbations of m-accretive operator, Nonlinear Anal. 16(1991), 771–780.

223. S. B. Nadler, Multi-valued contraction mappings, Pacific J. Math. 30(19 69), 475–487.

224. M. Nagumo, A theory of degree based on infinitesimal analysis, Amer. J. Math. 73(1951), 485–496.

225. M. Nagumo, Degree of mappings in convex linear topological spaces, Amer. J. Math. 73(1951), 497–511.

226. W. M. Ni, Some minimax principles and their applications in nonlinear elliptic equations, J. Anal. Math. 37(1980), 248–275.

227. L. Nirenberg, Topics in Nonlinear Functional Analysis, Lecture Notes, Courant Institute, New York, 1974.

228. L. Nirenberg, Variational and topological methods in nonlinear problems, Bull. Amer. Math. Soc. 4(1981), 267–302.

229. R. D. Nussbaum, Degree theory for local condensing maps, J. Math. Anal. Appl. 37(1972), 741–766.

230. R. D. Nussbaum, On the uniqueness of the topological degree for k-set contractions, Math. Z. 137(1974), 1–6.

231. R. D. Nussbaum, Estimates for the number of solutions of operator equations, Appl. Anal. 1(1971), 183–200.

232. V. V. Obkhovskii, On the topological degree for a class of noncompact multivalued mappings, Funkt. Anal. 23(1982), 82–93.

233. H. Okochi, On the existence of anti-periodic solutions to a nonlinear evolution equation associated with odd subdifferential operators, J. Funct. Anal. 91(1990), 246–258.

234. H. Okochi, On the existence of anti-periodic solutions to nonlinear parabolic equations in noncylindrical domains, Nonlinear Anal. 14(1990), 771–783.

235. T. Okon, A generalized topological degree in admissible linear spaces, Zeit. für Anal. Anwend. 14(1995), 469–496.

236. D. O'Regan and R. Precup, Theorems of Leray-Schauder Type and Applications, Gordon and Breach, 2001.

237. R. S. Palais, Morse theory on Hilbert manifolds, Topology 2(1963), 299–340.

238. R. S. Palais, Lusternik-Schnirelman theory on Banach manifolds, Topology 5(1966), 115–132.

239. R. S. Palais, Critical point theory and the minimax principle, Proc. Symp. Pure Math. 15(1970), 185–212.

240. R. S. Palais and S. Smale, A generalized Morse theory, Bull. Amer. Math. Soc. (1964), 165–171.

241. D. Pascali and S. Sburlan, Nonlinear Mappings of Monotone Type, Noordhoff, Leyden, 1978.

242. N. Pavel, Nonlinear Evolution Operators and Semigroups, Lect. Notes Math., 1260, Springer-Verlag, Berlin, 1987.

243. A. Pazy, Semigroups of Linear Operators and Applications to Partial Differential Equations, Springer-Verlag, Berlin, 1983.

244. W. V. Petryshyn, Antipodes theorems for A-proper mappings of the modified type (S) or $(S)_+$ and to mappings with the P_m property, J. Funct. Anal. 71 (1971), 165–211.

245. W. V. Petryshyn and P. M. Fitzpatrick, A degree theory, fixed point theorems, and mapping theorems for multivalued noncompact mappings, Trans. Amer. Math. Soc. 194(1974), 1–25.

246. W. V. Petryshyn, On the solvability of $x \in Tx + \lambda Fx$ in quasinormal cones with T and F k-set contractive, Nonlinear Anal. 5(1981), 585–591.

247. W. V. Petryshyn, Existence of fixed points of positive k-set contractive maps as consequences of suitable boundary conditions, J. London Math. Soc. 38(1988), 503–512

248. W. V. Petryshyn, Generalized Topological Degree and Semilinear Equations, Cambridge University Press, Cambridge, 1995

249. R. S. Phillips, Integration in convex linear topological spaces, Trans. Amer. Math. Soc. 47 (1940), 114–145.

250. T. Pruszko, Some applications of the topological degree theory to multivalued boundary value problems, Dissert. Math. 229(1984), 1–52.

251. T. Pruszko, Topological degree methods in multi-valued boundary value problems, Nonlinear Anal. 9(1981), 959–973.

252. P. H. Rabinowitz, Some aspects of nonlinear eigenvalue problems, Rocky Mountain J. Math. 3(1973), 161–202.

253. P. H. Rabinowitz, A note on topological degree for potential operators, J. Math. Anal. Appl. 51(1975), 483–492.

254. I. Rachunkova and S. Stanek, Topological degree method in functional boundary value problems, Nonlinear Anal. 27(1996), 153–166.

255. S. Reich, Some problems and results in fixed point theory, Contemporary Math. 21(1983), 179–187.

256 T. Riedrich, Die Räume $L^p(0,1)$ $(0 < p < 1)$ sind sulässig, Wiss. Z. Techn. Univ. Dresden 12(1963), 1149–1152.

257. T. Riedrich, Der Raum $S(0,1)$ ist zulässig, Wiss Z. TU Dresden 13(1964), 1–6.

258. J. W. Roberts, A compact convex set with no extreme points, Studia Math. 60(1977), 255–266.

259. R. T. Rockafellar, Characterization of the subdifferential of convex functions, Pacific J. Math. 17(1966), 497–510.

260. R. T. Rockafellar, Convex Analysis, Princeton University Press, Princeton, 1970.

261. R. T. Rockafellar, On the maximal monotonicity of subdifferential mappings, Pacific J. Math. 33(1970), 209–216.

262. R. T. Rockafellar, Local boundedness of nonlinear monotone operators, Michigan Math. J. 16(1970), 397–407.

263. S. Rolewicz, Metric Linear Spaces, Polish Scientific Publishers, Warszawa, 1972.

264. E. H. Rothe, A relation between the type numbers of a critical point and the index of the corresponding field of gradient vectors, Math. Nachr. 4(1950-51), 12–17.

265. E. H. Rothe, On the Cesari index and the Browder-Petryshyn degree, in International Symposium on Dynamic Systems, pp. 295–312, Academic Press, New York, 1977.

266. E. H. Rothe, Introduction to Various Aspects of Degree Theory in Banach Spaces, Amer. Math. Soc., Providence, RI, 1986.

267. W. Rudin, Functional Analysis, MacGraw-Hill, New York, 1973.

268. K. P. Rybakowski, On the homotopy index for the infinite-dimensional semiflows, Trans. Amer. Math. Soc. 269(1982), 351–382.

269. K. P. Rybakowski and E. Zehnder, On a Morse equation in Conley's index theory for semiflows on metric spaces, Ergodic Theory Dyn. Syst. 5(1985), 123–143.

270. B. Rzepecki, Remarks on Schauder's fixed point principle and its applications, Bull. Acad. Polon. Sci. Sér. Sci. Math. 27(1979), 473–480.

271. B. N. Sadovskii, Limit-compact and condensing operators, Uspekhi Mat. Nauk. 27(1972), 81–146.

272. S. S. Schaefer, Topological Vector Spaces, Macmillan, New York, 1966.

273. J. Schauder, Der Fixpunktsatz in Funktionalraumen, Studia Math. 2(19 30), 171–180.

274. J. T. Schwartz, Nonlinear Functional Analysis, Gordon and Breach, New York, 1969.

275. J. C. Scovel, A simple intuitive proof of a theorem in degree theory for gradient mappings, Proc. Amer. Math. Soc. 93(1985), 751–753.

276. V. M. Sehgal and E. Morrison, A fixed point theorem for mutifunctions, Proc. Amer. Math. Soc. 38(1973), 643–646.

277. H. W. Sieberg, Some historical remarks concerning degree theory, Amer. Math. Monthly (1981), 125–139.

278. M. Sion, On general minimax theorems, Pacific J. Math. 8(1958), 171–176.

279. I. V. Skrypnik, Nonlinear Elliptic Equations of Higher Order, Nauk. Dumka, Kiev, 1973.

280. I. V. Skrypnik, Sovlvability and properties of solutions of nonlinear elliptic equations, VINITI, Vol. 9, 131–254, Moscow, 1976.

281. M. Struwe, Variational Methods, 2nd ed., Springer-Verlag, Berlin, 2000.

282. A. Sulkin, Positive solutions to variational inequalities, a degree theoritic approach, J. Diff. Equat. 57(1985), 90–111.

283. J. X. Sun, A generalization of Guo's theorem and applications, J. Math. Anal. Appl. 126(1987), 12–17.

284. Y, Sun and J. X. Sun, Multiple positive fixed points of weakly inward mappings, J. Math. Anal. Appl. 148(1990), 431–439.

285. W. Takahashi, Nonlinear Functional Analysis, Yokohama Publishers, Japan, 2000.

286. J. F. Toland, A Leray-Schauder degree calculation leading to nonstandard global bifurcation results, Bull. London Math. Soc. 15(1983), 149–154.

287. S. L. Trojansky, On locally uniformly convex and differentiable norms in certain nonseparable Banach spaces, Studia Math. 37(1971), 173–180.

288. T. S. Tucker, Leray-Schauder theorem for P-compact operators and its consequences, J. Math. Anal. Appl. 23(1972), 355–364.

289. A. Tychonoff, Ein fixpunktsatz, Math. Ann. 111(1935), 767–776.

290. M. M. Vainberg, A new principle in the theory of nonlinear equations, Usphehi Mat. Nauk SSS 15(1960), 243–244.

291. M. Vath, Fixed point theorems and fixed point index for countably condensing maps, Topological Methods in Nonlinear Anal. 13(1999), 341–363.

292. M. Visser, Topological degree for supersymmetric chiral models, Phys. Rev. D. 32(1985), 510–512.

293. V. A. Volpert and A. I. Volpert, Properness and topological degree for general elliptic operators, AAA 3(2003), 129–181.

294. A. I. Volpert and V. A. Volpert, Application of the theory of the rotation of vector fields to the investigation of wave solutions of parabolic equations, Trans. Moscow Math. Soc. 1990 (1990), 59–108.

295. A. I. Volpert and V. A. Volpert, Construction of the Leray-Schauder degree for elliptic operators in unbounded domains, Ann. Inst. H. Poincaré Anal. Non Linéaire, 11(1994), 245–273.

296. V. A. Volpert, A. I. Volpert and J. F. Collet, Topological degree for elliptic operators in unbounded cylinders, Adv. Differential Equations 4(1999), 777–812.

297. Z. Q. Wang, Symmetries and the calculations of degree, Chinese Ann. Math. Ser. B 10(1989), 520–536.

298. G. F. Webb, Continuous nonlinear perturbations of linear accretive operators, J. Funct. Anal. 10(1972), 191–203.

299. J. R. L. Webb, A-properness and fixed points of weakly inward maps, J. London. Math. Soc. 27(1983), 141–149.

300. J. R. L. Webb, Topological degree and A-proper operators, Linear Algebra Appl. 84(1986), 227–242.

301. J. Weiberg, A topological degree of locally A-proper maps, J. Math. Anal. Appl. 65(1978), 66–79.

302. M. Willem, On a result of Rothe about the Cesari index and A-proper mappings, Boll. Un. Mat. Ital. 17(1980), 178–182.

303. M. Willem, Topology and semilinear equations ar resonance in Hilbert spaces, Nonlinear Anal. 5(1981), 517–524.

304. V. Williams, Closed Fredholm and semi-Fredholm operators, essential spectra and perturbations, J. Funct. Anal. 20(1975), 1–25.

305. S. F. Wong, The topological degree of A-proper maps, Canad. J. Math. 23(1971), 403–412.

306. S. F. Wong, A product formula for the degree of A-proper maps, J. Funct. Anal. 10(1972), 361–371.

307. J. Z. Xiao and X. H. Zhu, Topological degree theory and fixed point theorems in fuzzy normed spaces, Fuzzy Sets and Systems 147(2004), 437–452.

308. K. Yosida, Functional Analysis, Springer-Verlag, 1969.

309. T. Yuasa, Differential equations in a locally convex space via the measure of noncompactness, J. Math. Anal. Appl. 84(1981), 534–554.

310. E. H. Zarantonello, Solving functional equations by contractive averaging, Math. Research Center Report 160, Madison, WI, 1960.

311. E. H. Zarantonello, Projections on convex sets in Hilbert space and spectral theory, Contributions to Nonlinear Functional Analysis, Academic Press, 1971.

312. E. Zeidler, Nonlinear Functional Analysis and its Applications, Springer-Verlag, Berlin, 1990.

313. V. V. Zelokova, On a topological degree for Vietoris type multimaps in locally convex spaces, Sb. Stat. Asp. Stud. Mat. Fak. Vgu. (1999), Voronezh, 45–51.

314. Y. C. Zhao, On the topological degree for the sum of maximal monotone operator and generalized pseudomonotone operator, Chinese Ann. Math. 4(1983), 241–253.

Chapter 9

SUBJECT INDEX

L

M

N

O

P